SOAS Studies on South Asia

INDIA'S INDUSTRIAL CITIES

SOAS Studies on South Asia

India's Industrial Cities

Essays in Economy and Demography

Nigel Crook

DELHI

OXFORD UNIVERSITY PRESS

BOMBAY CALCUTTA MADRAS

1993

Oxford University Press, Walton Street, Oxford OX2 6DP

New York Toronto
Delhi Bombay Calcutta Madras Karachi
Kuala Lumpur Singapore Hong Kong Tokyo
Nairobi Dar es Salaam
Melbourne Auckland

and associates in
Berlin Ibadan

Printed at Chaman Offset Press, New Delhi 110002
and published by Neil O'Brien, Oxford University Press
YMCA Library Building, Jai Singh Road, New Delhi 110001

CONTENTS

ACKNOWLEDGEMENTS

I would like to be able to acknowledge by name all those who have helped form the ideas that have grown into the essays contained in this book: but that would be an impossible task and inevitably I have to be selective. It was C.R.Malaker who first prompted me to explore the rich source of data in the 1961 Census of India, on which several of the essays are based. Both Terry Byres and Pramit Chaudhuri made valuable comments and prompted further enquiry when several of the essays were at an early stage of development. Tim Dyson and Clare Becket kindly read and commented on Chapter 2. I also benefited greatly from the insights of Radhika Ramasubban on the subject matter of this chapter. Prabhat Patnaik, Heather Joshi, and Maire Ni Bhrolchain all contributed at various stages to my refinement of Chapter 5. An anonymous referee helped me considerably to tighten the argument in Chapter 6; a similar benefit was conferred on Chapter 7 by Peter Ayre, Francis Teal, and Lalit Deshpande. I owe a particular debt of gratitude for the knowledge and understanding I have gained from Meera Bapat without which Chapter 8 would never have been written. Bhanwar Singh facilitated the production of the opening and closing chapters. A second referee encouraged me to improve the style of data presentation, I hope to some effect. Finally I am indebted to Monisha Shah for undertaking the painstaking task of preparing a camera-ready copy.

At various points on the road to the completion of these essays my research has been facilitated by the hospitality and assistance provided by academic institutions: in particular I should like to mention the Gokhale Institute of Politics and Economics in Pune, the Indian Statistical Institute in Calcutta, and the School of Oriental and African Studies in London. I am also grateful to the SOAS Research and Publications Committee for a grant towards the preparation of the final text.

The shortcomings that remain should be attributed to me alone.

N.C.

Chapter 1

INTRODUCTION: THE URBANIZATION DEBATE, THE PREMISE AND THE THEME

Debate and Premise

As we slowly enter the environs of yet another city, many contrary thoughts will drift across the mind. Close to the railway lines, settlements crop up between the dusty vegetation, dwellings·often dilapidated, makeshift, but sometimes innovative and well-built. Colourfully clad women crouch to wash clothes in muddy pools of water. Buffalo wallow; hogs root around the rubbish. Naked children play, spin tops, fly kites. There is squalor and stench, but liveliness and livelihood. Then comes a market where people throng; rickshaws and bicycles clutter the approach to a railway crossing: the morning shift comes off, life goes on. Beyond the expanse of orange-tiled roofs rises a massive chimney, smoke drifting to thicken the afternoon; and, beyond the chimney, purple flat-topped hills. We grind our way into the station and a small boy climbs aboard; eyes misty and nose running, he sweeps the carriage floor and asks for coins. Had he stayed in the hills he would probably have died.

The industrial city is one of India's major resources. It is a creation of the process of industrialization, which is itself fundamental to economic growth, critical though one may be of the character that the growth process has assumed. The demand for industrial labour has been the prime mover behind the formation of these cities, and a significant factor in the sustained progress of urbanization itself. The demographic outcomes of industrialization are diverse, as diverse as are the industrial processes themselves. Their social implications are problematic: but these need addressing on their own terms, and not by stifling the very process itself. These are the themes, and this is the message, that run through the seven essays of this book.

There has been a pervasive tendency in India and elsewhere to view urbanization in a critical, if not hostile, light. Studies in the 1950s and 1960s attempted to explain manifestations of underemployment in urban areas in terms of the 'over-urbanization' of developing countries given their levels of industrialization (Davis and Golden, 1954): such a notion with its loose theoretical underpinning was adequately dismissed by Sovani (1964), and more recently rejected again, empirically, by Preston (1979). But a more carefully reasoned and widely discussed literature developed around the idea that excess labour supply to urban areas was induced by market imperfections that kept urban wages high in factory and government sectors (Todaro, 1969): the details and the implications of this thesis, more than the fundamental idea, have come in for considerable criticism. That the marginal returns to investment in industry were sustained at an artificially high level (in comparison with those to agriculture) was also the underlying feature of Lipton's understanding of the outcome he termed 'urban bias': this view has also provoked debate (Lipton, 1977). Unfortunately the diagnosis, whether correct or not, has been misinterpreted to imply that controls should be put on migration and urban growth; but not that more resources should be invested in agriculture. It remains the case that many writers, politicians, administrators, and middle class citizens generally, seem to subscribe to a view that the pace and pattern of urbanization in the developing countries generally, and in India no less, are undesirable. Much effort and less thought is devoted to devising ways of dispersing and decentralizing populations (Gugler, 1988; Hardoy and Satterthwaite, 1990).

This book hopes to redress the imbalance in the literature and debate. Whereas countless studies have come to dwell on the problem of labour supply (and the negative role of excess migration), few focus on the question of labour demand (and the positive role of necessary migration). Cities have come to be regarded solely as population agglomerations, with housing, transport, and general infrastructural problems to contend with. In an economics framework, many studies have stressed their function as commercial and administrative centres, rather than as manufacturing foci. The essays herein introduce a countervailing tendency, which is to stress the industrial nature of urbanization, and to take as case-studies cities which have a predominantly industrial character. For cities are a necessary part of the industrialization process; they are created by it, and in turn they sustain it. These are facts that seem to have got lost in the literature and opinion hostile to urbanization.

Those who argue that urbanization is in some ways dysfunctional, that as a process it has become disjointed from the process of industrialization, and irrelevant if not harmful to the process of economic growth itself, have to confront one uncomfortable fact. In India since the 1930s the rate of urbanization has continued to increase decade by decade. If urban growth were detrimental to economic growth in the Indian capitalist system, surely it would have slowed in recent decades? But the evidence is that it has not. Neither capital nor labour will continue to be attracted in substantial quantities to a locality that cannot sustain a growth in per capita income. Take, for example a city such as Bombay which has several times been described as close to economic and social collapse since the initial steps were made to plan a satellite city across the Thana creek in the early 1970s; despite which Bombay has sustained a growth of around 40 per cent a decade for over thirty years. This would seem to represent an economic equilibrium, not a crisis. On the other hand, a city such as Sholapur, about which rather less concern has been expressed on paper, has attracted very little capital or labour since the 1940s. Net in-migration has been close to nil, and growth has barely exceeded that of natural increase.

In contrast, various policy attempts to stimulate the decentralization of industrial manufacture, the growth of small urban centres, with the expectation also perhaps of arresting the rate of urbanization itself, have met with limited success. They are not it seems conducive to the capitalist system of industrialization. Despite this fact intellectual and practical efforts still seem to go into trying to devise strategies and policies to make this alternative and ahistorical process work. Partly as a result of these prejudices and presumptions, rather little research has been devoted to trying to understand, on the other hand, why urbanization, and in particular large cities, are conducive to industrialization, and what are the tangible, and preferably measurable, benefits they confer. The production function analysis that is adopted by economists regards the quantity of industrial output as the product of components such as labour, capital, and land. It would be hypothesized that these components combine more effectively in large urban environments, resulting in more output for the same amount of inputs; technically speaking, the production function has a shift factor which is itself a function of the scale of urban agglomeration. The theoretical arguments in support of this have been developed quite well. They relate to the ready creation of labour markets in cities, for instance, which speed up the hiring and utilizing of appropriately skilled labour; to the fuller utilization and hence more adequate provision of social infrastructure like roads; and to the better communication between entrepreneurs in proximity to one another, thus enhancing capacity utilization in individual manufacturing concerns. But quantitative empirical studies there are few (Mera, 1973; Moomaw, 1981; Shukla, 1988). Nor do the essays in this volume seek principally to address this important question: that is more an agenda for further research. These observations, however, serve to point out that the same positive stance towards urbanization and its relationship with industrialization that is adopted herein has not been rejected either theoretically or empirically. One might be surprised to learn this in view of the clamour in the anti-urban lobby.

If the urbanization literature has suffered from inadequate information and research, it has suffered even more from misdiagnosis. Whatever the economic benefits of the process, the malaise that results is keenly observed. Underemployment, slum settlements, congestion and pollution meet the urban dweller's eye. To attribute these manifestations of population agglomeration to urbanization per se is questionable, as we shall show in the later essays of this volume. It also reflects a tendency to confine analysis to the large and growing city, a bias that the essays herein will make some effort not to replicate. To attempt to mitigate these problems by stifling urban growth itself is indicative of the misdiagnosis; at worst it is to put the patient to death rather than attending to the wounds.

It should not be concluded from the above discussion that the social and economic problems arising in the process of urbanization will go undiscussed in what is to follow. In fact, on the contrary, our focusing on some of the new industrial towns and cities, and giving them a prominence they have not usually received in the literature, has been prompted by the need to bring to the fore these very problems in places which the literature seems largely to have overlooked. The industrial new-towns, especially the steel cities, may be felt by the reader to dominate the pages of this book. This is partly also to redress an imbalance in studies of, and policies towards, urban economic problems. For the preoccupation with the mature and metropolitan cities like Calcutta and Madras has served to obscure the fact that the industrial new-towns like Durgapur and Bhilainagar are now forty years old, and might also be developing symptoms of decay. The existence, causes and consequences of the latter are discussed in the earlier essays in this volume.

The Essays and their Theme

Viewed as a process, the relationship between industrialization and the formation of cities starts with the industrial demand for labour. In this sense Chapter 3, on the recruitment and migration of the industrial labour force, is the cornerstone of this work, and the distillation of its central theme. Labour demand is the prime mover of industrial urbanization. Labour demand is, however, a complex factor, which gives rise to a diversity and complexity in the migration process; that diversity in itself becomes a theme in this and the other essays herein. Labour is differentiated by characteristics such as age and sex, in part according to the needs of the industrial employer; and the consequent characteristics of the demographic agglomeration will come to reflect this fact. Furthermore, ethnic, linguistic, and caste composition of a city can be traced to recruitment policy and the derived demand for labour as the local economy develops.

Much of the detailed data on which this study is based come from the 1961 census, which presents a 'still' picture in a process which was developing at the high tide of the Second Five Year Plan, India's major industrialization strategy of the 1950s. The complex demographic cross-currents can only be understood fully by tracing the process back through the century. The evolution of that picture, the rise and the fall of cities, the interaction of their birth and death rates with earlier phases of recruitment and migration, is an historical process which we try to pursue from the date of the first major census in 1881: this date lies close to the first phase of factory industrialization in colonial India. The analysis which comprises Chapter 2 raises some intriguing questions on the historical role of the death rate in allowing the cities to grow. But its emphasis remains on the diversity and dynamics of the demographic experience in what is too readily referred to as if it were an undifferentiated process, namely just 'India's urbanization'.

Sometimes it helps to understand a process such as that of economic-demographic interaction if one can abstract from the factors that complicate the issue and cloud the analysis. This was our reason for taking the steel industry as a case-study in which to explore the recruitment, migration, demographic formation, and demographic evolution process in Chapter 4; a second reason was to explore the social implications of India's new-town development strategy, and to chart the emergence of a demographic and economic duality within the populations in these initially prestigious towns. The steel town phenomenon is, in part, a law unto itself. Arguably the nature of steel-making technology determines a unique demography, both in time and space. But the chapter proceeds to relax the controls, to examine the effect of scale economies in industries other than steel, and hence to enrich the explanation of the diversity in demographic dynamics that are witnessed, as industry develops new technical and economic characteristics down to the 1990s. Some of these changes need to be situated in the framework of the changing political economy itself.

Chapter 5 presents a case of a different kind for the study of social and demographic dynamics. For reasons of comparative advantage some locations have attracted industrial investments in a single industry only. The resulting urban formations, be they textile, mining, or steel towns have failed to diversify their industrial base over time. Questions arise here concerning their vulnerability to economic shocks, to disinvestment, or to the failure of investment to grow, and the economic strategies open to their urban populations to mitigate the consequences. In this chapter we have reconstructed the empirical evidence from secondary data to throw light on the potential for household survival, through family linkages, resource tranfers, and household economic diversification: this also involves us in an analysis of the demography of household formation in the towns. As in the previous chapter the case illustrates clearly the potential for industrial investment behaviour and endogenous demographic dynamics to fall into disequilibrium at the level of the individual city. Cycles of production and reproduction fail to coincide. A case can be made here for social intervention in investment planning to prevent the perpetuation of these 'industrial monocultures'.

If industry needs labour, and acquires it initially through migration, it also needs the reproduction of labour, both in terms of day to day survival and replacement of energy loss, and in terms of demographic reproduction so that as one generation retires another enters the industrial fray. Cities in history are notorious for consuming men and women, so that major industrial expansion would have faced a labour constraint; more recently a remarkable reversal has occurred with labour becoming plentiful enough to provide the growth in recruitment and reserves that industry needs. Nevertheless, the question of reproducing quality labour continues to arise. Chapter 6 interprets population and health policies in this light. It explains growing state intervention in the environmental control of large cities in terms of consumption externalities and the rising middle-class in India's urban society; and indicates in terms of the costs of environmental control and aspirations towards middle-class life-styles, how, in urban areas at least, the needs of the state and of the working class family coincide on the question of fertility decline.

As we observed earlier, urbanization, while fulfilling the requirements of an industrializing economy, brings with it certain costs. Chapter 7 addresses itself to the problem of meeting those costs in ways that are both efficient and equitable. It tries to show in detail that many of the policies often proposed would retard or redistribute the growth of cities thereby suppressing both their social benefits and their social costs; in some cases the actual cost-incurring activity would not as a result be restrained at all. It is not migration to cities that causes water sources to be over-used: it is the failure of the state to supply sufficient water and recover the cost from the users. It is not the location of industry in cities that causes air pollution; it is the emission of smoke. This is what we mean by a mis-diagnosis of the problem. As policy issues these are very much alive in the debate of the 1990s, and as political issues very much a cause of conflict.

The one area of urban economic management that affects or afflicts the lives of urban citizens most acutely is the housing question. Again a failure to grasp the nature of the problem has led to impractical administrative solutions. Again there is political conflict, and for a time in the 1970s it appeared that demolition of squatter settlements, a tactic that amounted to a form of forcible migration control, would prevail. In the 1980s, for reasons probably rooted in the political economy of the state, there is an incipient tendency towards more constructive outcomes. Whatever the political forces that hold sway, Chapter 8 shows how certain technical and administrative solutions that have been widely canvassed are economically impossible.

It may seem from the foregoing that too much emphasis is being placed on the economic efficiency of a process, and too little on matters of distribution and equity. The author's first micro-study was conducted in seven shanty settlements in Poona. Two of these were peopled by migrants who had fled to the city as a refuge in the second year of drought in the Marathwada region of the State. We followed up these migrants four years later (Bapat, Crook and Malaker, 1989), and they had survived the problems of coming to grips with an already congested city and labour market; had they remained in their villages they would not have survived. The better solution to their situation would have been a greater allocation of resources to the Marathwada region and a greater efficiency in their use to ensure the region survive the drought. But it would have been no solution to have restricted migration of the people to Poona, or to have restricted employment creation in Poona. It is also in the spirit of this understanding that the following essays are written.

Chapter 2

INDIA'S URBAN DEMOGRAPHY IN HISTORICAL PERSPECTIVE

Introduction

It is time to write a demography of India's towns and cities. Already too many erroneous beliefs, or at least controversial ideas, have become almost accepted orthodoxy; too many generalizations about India's urbanization have been allowed to obscure the diversity of the process, a diversity that is both regional and typological. Above all what is lacking is a sense of the *dynamics* of the process: for even in the current century there have been periods of urban stagnation as well as rapid growth, and while the pattern of urban demography revealed in the recent censuses may be analysed in aggregate, what is going on beneath the surface is the interplay of a myriad of cross-currents as, relative to one another, individual towns and cities rise and fall.

In 1899 Adna Ferrin Weber's study, *The Growth of Cities in the Nineteenth Century*, was published (Weber, 1963). It was subtitled *A study in statistics*, and, while not being in any way a treatise in technical demography, it remains in my view the outstanding empirical study of the demography of urbanization. It raises nearly every question of interest to the urban demographer, and many of the questions that concern the urban economist also. Perhaps one of its strongest points is an acute awareness of a diversity of experience that deserves the attempt to identify patterns relating to typology and region: it should be no accident that the study was entitled *the growth of cities* rather than *urbanization*, emphasizing the sum of parts rather than the undifferentiated whole. The present essay will present an overview of India's urban demography in the light of issues of interpretation and historical experience raised in Weber's classic work.

Growth, Migration, and Mortality

Towns and cities grow in demographic terms through the excess of births over deaths, through the excess of in-migrants over out-migrants, and through the incorporation of rural population as urban boundaries get redrawn. In the nineteenth century European city, growth depended decisively on the excess of migration over mortality and for many urban areas over considerable periods there was no such excess. As Weber indicated from Swedish statistics that were exceptionally good, the real explanation of city growth in the last half of the nineteenth century was to be found in the natural increase, that is to say from the fall in the death rate. In many European cities in-migration was needed to prevent natural decrease. Hence the proportional contribution that migration makes towards urban growth is a multi-faceted concept, and often misleadingly quoted in isolation. For example, in-migration is more likely to make a major contribution if urban death rates are high or birth rates are low. The former was the case in nineteenth century Europe; the latter will become the case if fertility continues to fall in India. But during the time when death rates have fallen substantially but birth rates have fallen less, migration will usually make only a minor contribution to urban growth. This has been generally the case in post-Independence India. Another manifestation of that fact is that the proportion of the population classified as urban is only edging up gradually from census to census. As in the rural areas, so in the urban areas population growth is due to substantial excess of births over deaths, and the main reason for that is the relatively low death rates. To put the point graphically, given current levels of fertility and mortality, even if migration dropped to zero, Calcutta, Bombay, Delhi, and Madras would each increase by one half within the next twenty years as a result of natural increase alone. This all points to one conclusion: it is absurd to imagine that the so-called urban problem would vanish if migration could be controlled. Large urban populations with a natural growth potential already exist.

This discussion invites a question on another demographic aspect of urbanization, which has not, I think, been adequately posed by demographers before. Was it the case that in India, as in Europe and America, urban deaths exceeded births in the ninteenth century? Further to this, given that the sanitary revolution that is supposed to have contributed substantially to the fall in the European urban death rate has failed to touch a substantial proportion of India's urban population even today, when and why did urban deaths in India first fall below urban births? When indeed did the urban death rate first fall below the rural? There is a clear margin between the two today with urban death rates standing at ten per thousand or less and rural at fifteen per thousand or so, a difference of at least five per thousand; but how far back this century did a difference of this order obtain?

We said above that we do not have accurate figures for birth or death rates for Indian cities before the 1970s. But what we do have are urban growth rates (that is natural increase plus net migration), and it is possible to compare these with those of the total population (which was between 80 per cent and 90 per cent rural throughout the pre-Independence period). Table 2.1 does this for each census decade from 1881 to 1981. During the first of these decades, 1881-91, urban growth rates exceeded rural-plus-urban growth rates (total growth rates) by only a small margin, 0.2 per cent per annum. Total growth rates were quite considerable, indicating a gap between birth and death rates; most of that gap will have been due to more favourable death rates than were to follow for three decades. This in turn may have been accompanied by a reduced need for migration of the destitute and starving to the towns where work or food would be expected to be more available, hence part of the explanation for the slow urban growth. Alternatively, the slow urban growth may have reflected higher mortality in the urban areas themselves, for reasons to be addressed shortly. With a reduced growth in the following decade (1891-1901) reflecting higher mortality as famine returned to the land, increased rural-urban migration widened the gap between the total and urban growth rates, though to only 0.5 per cent per annum. However, in the next decade (1901-11) total growth rates actually exceeded urban growth rates. It was possible that out-migration from the cities exceeding in-migration was responsible; but higher mortality in the urban areas is also plausible, and indeed the former may be related to the latter as people flee the city to escape epidemic disease. There is a simple explanation for high urban mortality between 1901 and 1911, and that is the plague (which entered India in 1896). The cause of death statistics available and observations at the time seem to suggest that urban environments initially favoured plague. Plentiful waste matter (to feed the rats), high density living (to speed the progress of fleas), and insufficient water to maintain the cleanliness of people, clothes and bedding characterise urban areas. In these environmental respects rural areas were probably marginally better, and especially in avoiding concentrations of waste matter. Bombay's demographic history vividly illustrates the case: the crude death rate as registered rose from around 20 to 30 per thousand prior to 1895 to between 40 and 65 per thousand from 1896 to 1910.[1] After 1896 there was out-migration to escape the disease.[2] Benares, Delhi, Ahmedabad and Lahore also registered death rates of between 50 and 60 per thousand, in 1907 and 1908. From 1891 to 1901 twelve of the so-called class I cities (i.e.

those of 100,000 population and more) suffered a decline in population along with Bombay (Table 2.1). From 1901 to 1911 fourteen of these cities lost population, and the average rate of change of all the class II cities (which have between 50,000 and 100,000 population) was negative also. Between 1911 and 1921 ten class I cities declined (the influenza pandemic being a contributory factor here). It seems highly likely that most of these large cities in decline were enduring an excess of deaths over births for which in-migration could not compensate (in conjunction probably with some out-migration to escape the horrors of disease). The parallel with many European cities as high mortality foci during certain decades of the nineteenth century is plausible.

The contrast in growth rates between cities is also striking, a fact brought out by Weber who devoted a few pages to an analysis of the Indian censuses of 1881 and 1891. The population in the whole of Bombay Presidency (currently part of Maharashtra and Gujarat) grew at 1.3 per cent per annum, whereas the cities of Bombay, Poona, Ahmedabad, and Surat managed an aggregate growth of only 0.8 per cent. Amritsar, Jubbulpore and Patna in the north went into decline. Madras City in the south grew more slowly than Madras province. By contrast there were a few rapidly industrializing towns during the same period which managed a rate of in-migration that could well outdo any losses due to high mortality (if losses there were in their case): for example, Hubli with a growth of 3.6 per cent annually, Karachi (3.6 per cent also), and Ajmer (3.5 per cent).

The general picture then seems to be one of cities as death traps: 'it is almost everywhere true that people die more rapidly in cities than in rural districts' wrote Weber (p. 343). Today we know that Weber's summary no longer holds for India, nor indeed for the rest of Asia, nor in Africa or Latin America. But when and why did the change in India take place? A glance at Table 2.1 might suggest that from 1921 onwards urban death rates began to fall below the rural. But at the same time in-migration must have picked up also; for the years between the first and second World Wars witnessed some industrial growth, albeit slow by present day standards. By the time the Sample Registration System became operative in the late 1960s all the states of India were experiencing urban death rates that were lower than the corresponding rural rates. In 1981, when child mortality data were collected at the census (on a sample basis) for the first time, the urban advantage of lower mortality held for almost every one of the 300-odd districts also: this means that small or medium sized towns were sharing in this experience (GOI, Registrar General, 1988).[3] Furthermore, these latter data are not vitiated by the need to allow for the more favourable age composition of urban areas.

There is as it happens a series of recorded deaths for the Central Provinces where registration is believed to be substantially complete. The rural and urban data are available separately, and the demographer Tim Dyson has put together (without demographic manipulation) a run of crude death rates from the 1870s through to the 1940s.[4] In the 1870s urban death rates are clearly higher than rural; from the early 1880s until the mid-1890s the reverse is the case, and Dyson has suggested that this reflects the selective impact of the famine years; it also corresponds, however, with some individual years of unusually light urban mortality, i.e. death rates under 30 per thousand. From the mid 1890s the urban death rates rise again above the rural, and remain above until the late 1920s or early 1930s, when a reversal begins to take place. In a few of the years the differential is substantial, at around 100 per cent; these are undoubtedly years of plague epidemics. By contrast the influenza epidemic caused a 35 per cent excess of rural over urban death rates. Table 2.2 presents Dyson's series for the census years only, together with SRS data for recent years. This series generally accords well with the periodization we offer above. Differential improvement or deterioration in registration between urban and rural areas cannot of course be ruled out as a contributory explanation, nor can changing age structures; and the denominator can only be measured accurately for census years. But the totality of the evidence is strongly suggestive of real social and demographic changes taking place.

The question that arises is what factors determining mortality decline could have operated disproportionately well in towns and cities? This is a complex problem and it would be unrealistic to expect a simple solution. From around 1920 the answer might be the gradual disappearance of the disease that operated particularly successfully in urban conditions, namely plague. Plague had disappeared almost totally by the 1950s without availability of medical cure. It is tempting to assume that the causes of its disappearance were neither social nor environmental, as cholera failed to decline in the same way. Plague accounted for a million recorded deaths in India in epidemic years at the start of the century; by the 1930s it had come to account for under 100,000. In some years cholera took fewer lives at the turn of the century than did plague, but at least twice as many by the 1930s (GOI, Sanitary Commissioner).[5]

In the case of Europe and America in the nineteenth century whatever advantage the urban areas might have possessed in terms of social composition and concentration of medical facilities was significantly offset by their deleterious physical environments. Improvements in the latter began to occur in the late nineteenth century; such improvements had not come to benefit the working classes until the beginning of the current century.[6] Weber observed that the western world's metropolitan cities did not have the highest mortality among urban localities at the turn of the century; in fact it was close to the lowest. This holds true in India today. The 1981 Census survey shows child mortality to be lower in Bombay and Calcutta than in most of the urban areas of their respective states.[7] Weber saw the effects of improving health in the cities manifest in London's death rate falling from 50 to 25 per thousand over a 25 year period prior to the end of the nineteenth century. 'In the introduction of enlightened sanitary methods we should expect the largest cities to lead the way,' he wrote. This may well have accounted for the nineteenth century experience of the English population taken as a whole. In India the sanitary 'revolution was more of a 'gradation', and even in the metropolitan cities today it has a long way to go. It is however true that recent improvements in sanitary infrastructure, usually made with international financial support, have indeed been concentrated in the metropolitan cities. This is not to deny that the authorities were aware of the need for reform as colonial sanitary reports make abundantly clear from the 1860s onwards.[8] But the progress that was made lagged severely behind the need (probably more so than in England). Bombay is a good case in point, which also illustrates the problem of poorly co-ordinated planning of infrastructural improvements. The Tansa Lake water supply project was opened in 1892, but the city's drainage was quite inadequate to carry away the huge volume of water. In 1900, a year of famine that no doubt drove people to the city, there was a cholera epidemic too. The registered death rate reached 96 per thousand: it is clear that the water supply improvements in themselves were quite inadequate to prevent a catastrophe. The underlying problem of inadequate drainage is also referred to in the sanitary reports on Calcutta. As far as the non-metropolitan towns are concerned, sanitary infrastructure and housing remain in an abysmal state to this day, as an anecdote will illustrate: asked to survey the slum areas of a town in Uttar Pradesh, a research assistant wrote back to her professor: 'please give guidance: which areas of the town am I to suppose not to be a slum?' In the 1920s the authorities had commented on the effects of financial stringency on urban improvement: 'during the last ten years it is true much has been done to improve the sanitation of *large towns* [italics mine]; but financial stringency has of late been responsible for failure to maintain progress' (British Government, House of Commons, 1924-5).

To understand the political factors underlying such tardy progress would require a treatise in itself. Here we can go no further than demographic observations. But in general our case is that urban environmental conditions in India often resulted in higher mortality than in rural areas in the past, exactly as in Europe. In the course of the present century urban mortality declined, until today it is lower than that in rural areas. A partial factor in the explanation for this was the arrival and disappearance of plague, which seem both to have been exogenously determined.

The author knows of no study that has rigorously explained contemporary rural-urban differentials in mortality. Demographers would point to the age composition, which tends to favour lower urban death rates, but age-standardized measures still show a differential in favour of the towns. There may also be a self-selection effect among migrants to urban areas favouring healthier adults. And, more recently, lower urban fertility and wider birth-spacing will have contributed to lower child and infant mortality (GOI, Registrar General, 1979).

The easier access to medical services in urban areas is of undoubted importance: availability of cure will offset the deleterious effects of unwholesome physical environments. Even in the poorest areas of major cities clinics and chemists have proliferated in the last twenty years. A study undertaken by the author and colleagues indicated the ready recourse of the poorest households to 'western' medicine in Bombay. Casual observation suggests that the picture is not very different in the non-metropolitan cities either (Crook, Ramasubban and Singh, 1991).

The importance of social composition also needs serious consideration in our search to explain the lower urban mortality. Studies of nutritional status have shown similarly poor levels in urban as in rural areas among similar social classes (K.S.Jaya Rao, 1985). Studies of mortality have suggested much the same (Crook and Malaker, 1988). One might be tempted to ascribe most of the favourable mortality differential to social composition, i.e. a higher proportion of middle and upper income groups. But this would seem to run counter to the poverty studies of the 1960s and 1970s suggesting that no smaller a proportion of the population were below the poverty line in urban than in rural areas (evidence summarized in Mills and Becker, 1986). So all said, social composition probably goes some way (perhaps especially in the metropolitan cities), but not all the way, towards explaining lower urban than rural mortality in India, at least since Independence. In the last section of this essay we reflect on the importance of the development of literacy differentials between urban and rural populations. We have emphasised, in another study, the powerful interaction between social composition and urban amenity, manifest today for example in infant mortality differences within Bombay: around 50 per thousand in the salubrious and high-income wards in contrast with around 80 per thousand in the poorer and ill-serviced wards of the city (Crook, Ramasubban and Singh, 1991).

Age Structure, Migration and Fertility

Weber observed that crude birth rates were higher in the city than in the country, a fact he put down mainly to the younger age composition. In America and Europe it was more typical for women to migrate into towns than is the case in India today: in fact the male-dominated sex ratio in most Indian cities has led some scholars to argue that the overall affect is to depress the birth rate. As we shall further illustrate in later essays, these broad generalizations lose the intricacies of the undercurrents, which relate closely to dynamics and typology. For instance, the heavy industrial towns, that are largely the product of post-Independence planning, initially experience strongly male dominated in-migration, depressing the birth rate despite the young age structure of the migrants. After a lag of a few years young women join the men; at the same time male in-migration slows down. The combined effect is a tendency to equalise the sex ratio and push up the birth rate. At a single point in time sex ratios are not a good guide to family formation potential. They are merely symptomatic of a stage in the process. Lighter industrial towns, on the other hand, are often characterised by more balanced sex ratios, with less of a tendency for these to vary as the town develops. This is partly because the in-migration rate is likely to be more constant. Two contrasting scenarios will vivify the point. If a textile town, consisting of many small factories, adds a new factory each year, the annual flow of migrants to supply the workforce will be only a few hundreds. Whereas if a steel town consisting of one large plant adds a new blast-furnace and rolling mill every decade, the influx of migrants will be several thousand concentrated in one year out of ten. The tide will come in in waves, so to speak, rather than smoothly, even though at the end of the day the growth (or height of the tide) may have been the same in each case. In India most towns and cities have had the smoother growth profile, but because of the peculiar demographic interest of the other type, which have been especially concentrated in the east of the country, another essay in this collection is specifically devoted to their analysis. A further example of diversity is illustrated by those towns whose comparative economic advantage has been lost (often as a result of the exhaustion or lost competitiveness of an adjacent raw material). The result is little or no net in-migration, and the sex composition and eventually the age composition begin to parallel that of the rural population; the birth rates approximate those in the rural area. Such towns are also a minority in India, but they have come to exist (as in England) following a century of industrialization. The census gives a snapshot of the urban population, but, to continue the earlier analogy, it is not like a snapshot of the desert; it is more like a snapshot of the sea—it fails to capture the diverse and complex movements.[9]

We should not, however, get the impression that industrial typology is the sole determinant of the sex ratio of towns and cities, nor even of the sex ratio of migrants. Women are freer to move in some cultures than in others. Within India there is a striking difference between the mobility of women in the north and women in the south. Currently the sex imbalance of urbanward migrants in the major states ranges from about three men to every two women (in Bihar, Orissa, and Madhya Pradesh), to a ratio of one to one (in Tamil Nadu, Andhra Pradesh and Kerala; see table 2.3). One might argue that this reflects the demand for male-only labour in the north and centre of the country where most of the heavier industry is located, in comparison with the lighter industrial structure of the south, favouring female labour. An examination of individual occupations reported in another essay in this volume indicates that this is not the sole explanation (see Chapter 3). Furthermore, the main purpose of female migration is not for employment, nor even education, but for marriage or to accompany a spouse or parent. Altogether it is difficult to escape the conclusion that the sex profiles of urban migration for each state are in part a reflection of the relative social acceptability of female mobility. The north-south divide is consistent with what scholars have already told us about relative female autonomy, namely that the southerners are more liberated (Karve, 1965; Dyson and Moore, 1983).[10] What is striking is that this sex ratio characteristic, manifested as an urban phenomenon, has apparently been the case for a long time. For the sex ratios of the cities in 1911 indicate the same north-south divide: Calcutta, Bombay and Lahore had male-to-female ratios of at least three to two, whereas in Madras, Mysore and Hyderabad they were closer to one to one (Table 2.4). It seems therefore that for reasons of culture and employment composition some cities will have higher female-to-male sex ratios than others among the adult populations, which will in turn be favourable to higher birth rates, other things being equal. For most of their demographic experience, towns whose industries are light rather than heavy, and those located in southern India, have had higher female-to-male sex ratios.

However, the women in such towns may not in fact have the same propensity to marry nor the same marital fertility as those in other towns. Indeed the overall effect of later marriage and lower fertility is to reduce the birth rate in urban areas of India to levels below the rural in nearly every state.[11] Both women and men marry later in urban areas. Towns and cities (and especially the latter) might be expected to have a different social composition from rural areas: a proportionately larger middle-class associated with administrative services and the management and ownership of the developing commerce and industry, than the middle-peasantry in the countryside. They have also a more educated population, *across all classes*. Both characteristics go with delayed marriage. In some cases a selection effect among migrants may be at work: those towns whose migrants consist partly of women already working in the labour force or whose migrants are from cultural backgrounds where marriage is traditionally late will have an increased proportion of late marriers in their population. But the matter is complicated by the fact that the urban social environment seems to be particularly conducive to *changes* in fertility behaviour, and maybe also in marriage behaviour. Combinations of effects ranging from crowded living conditions to aspirations for elevated life-styles, which are in evidence everywhere, seem to encourage a move towards lower fertility across all social classes: these effects are concentrated in towns and particularly in the larger cities. Furthermore the speed with which new behavioural norms get communicated is faster in urban environments. In many of these respects it is doubtful whether anything has changed down the ages. Weber cites a quotation from the 1760s: 'more wants and increased splendour, with higher prices for the necessaries of life, keep men from marrying in the cities.' He is also conscious of the differentials relating to industrial and commercial typology and the concomitant social differentiation:

Dr. Ernst Engel was probably the first statistician to advance statistical data in favour of the proposition that it is chiefly the occupation rather than the mere association of peoples in large or small dwelling centres which causes the difference in fertility in city and country women... English statisticians, moreover, long since pointed out the high birth rate peculiar to mining and industrial districts. In Germany the cities which indisputably have the largest proportion of births to child-bearing women are the purely industrial cities of the Rhine-Westphalian district. On the other hand, the commercial cities with their greater wealth, comfort and culture have the lowest birth rate.

What is missing in this account is the sense that such characteristics change for individual towns and cities over time: urban demography is dynamic, urban social and population structure can be peculiarly unstable at times.

Urbanization, Education, and the Structure of the Labour Force

The focus of this essay has not been on the economic forces that have driven India's urbanization, but rather on the demographic outcome. Nor have we said much on the social composition of the towns and cities— except to suggest that there is much diversity and much change over time. One social aspect deserving of greater research is literacy. The rural-urban gap in the 1980s level of literacy is substantial. Clearly in part it reflects the contrasting economic composition of the two populations; but it also indicates the better access of even the poorer urban classes to schooling. The manifold effects of schooling are believed to interact favourably with infrastructure and services designed to promote health, lower mortality, delay marriage, and reduce fertility. Levels of female literacy are thought to be particularly sensitive indicators of likely progress in these respects. What is striking is the progress made in literacy, in the large cities at least, early in the present century (Table 2.5).

To take two examples for which I will present data on female literacy in the population aged 15 and above, in all Bengal the proportion literate crept up from 0.4 per cent in 1891 to 3.4 per cent by 1931, a very small improvement of only 3 per cent in absolute terms; but in the city of Calcutta, the 1891 proportion was 7.9 per cent (also quite small), rising, however, to 33.1 per cent by 1931, which in absolute terms is a substantial improvement of 25 per cent. Even in the United Provinces where female social progress was minimal (as indeed it remains today), such that the literacy was only 1.2 per cent in 1931, the level attained in Allahabad was 15.8 per cent, again a substantial rural-urban difference. The more favourable urban mortality that can be seen from the 1930s onwards was no doubt assisted by the admittedly limited social progress of this kind. I am not commenting on the desirability or necessity of the emergence of such rural-urban disparities in the course of economic development, but rather pointing to the fact that from quite early in the century the cities could in some way be regarded as progressive social foci in the land. We have suggested in this chapter that access to education, and later, access to modern medicine have benefited the poor in urban areas far more than the rural poor. These factors and their interaction may have gone some way towards explaining the lower fertility (and later the lower mortality) that emerged in the towns and cities.

Whereas educational composition has changed dramatically over the century, economists have drawn attention to the stagnant structure of the Indian labour force, with about 70 per cent employed principally in agricultural activities according to the 1901 and 1981 censuses (and this despite a substantial decline in the share of GDP originating in agriculture). This has to be contrasted with the shift in the population classified as urban from 11 per cent to 25 per cent over the same period.[12] We know that in 1981 about ten per cent of the urban workforce was engaged in agricultural activity, compared with 80 per cent of the rural workforce. Clearly the economic structure of the urban or rural workforces must have changed over the century (or both may have changed), otherwise we cannot explain increasing urbanization without decreasing proportions of the workforce in agriculture. Unfortunately the censuses do not readily allow us to discover exactly what has happened. However, we can surmise as follows. At the turn of the century there must have been a lower proportion of the 'rural' population engaged principally in agriculture than there is today; at the same time there may have been a lower proportion of the 'urban' population similarly engaged (see Appendix B). That is to say, as the century progressed, the population classified as rural became more agricultural; we may surmise that rural industry became increasingly urbanised. This does not necessarily mean that rural industry closed down, competed out of existence by urban-based industry, hastening the pace of rural to urban migration, though this must have happened to some extent especially with the increased commercialization and acceleration in urbanization in the 1930s. But it must also have happened that villages, or clusters of villages, and especially those on the outskirts of towns, became classified as urban as both boundaries and criteria for classification were redrawn. The process can be easily documented in the recent censuses when rapidly growing new towns are constantly engulfing neighbouring villages (and in the 1961 census especially, when the criteria themselves were changed). For this process to account for the progressive agriculturalization of the rural (and urban) workforces we must postulate that non-agricultural activities, including small-scale industrial production, tended to concentrate in particular localities, or to locate in the vicinity of existing towns and cities; this tendency has indeed been documented for more recent times. Insofar as more than a change in classificatory criteria is involved, the process reflects dynamic change: despite the apparent rigidity in the occupational structure of the labour force throughout the century, the social environment of the labour force was changing, especially from the 1930s onwards. The non-agricultural workforce was becoming increasingly urbanized.

NOTES

[1] It is worth noting that the crude death rates, and the infant mortality rates, are always higher in the City of Bombay around the turn of the century than in the Presidency in which the City lies (Municipal Commissioner of Bombay). One might argue that registration would be more complete in the City; but the crude birth rates, on the other hand, are higher in the Presidency than in the City as one might expect, suggesting that differential completeness of registration is not an important problem.

[2] Arnold, 1987.

[3] The 1972 Fertility Survey (GOI, Registrar General, 1976) collected information on children ever born and children surviving and tabulated the results for urban and rural areas of each state. In the majority of states the urban survival was better than the rural: only four were out of line, which may have been a reflection on data quality (two were quite small), or may indicate the more recent decline in urban mortality in these states revealed in these retrospective data.

[4] I am much indebted to Tim Dyson for generously giving me access to his data at this point. For a periodization of India's mortality regimes see Dyson (1989), which contains an eloquent illustration from Central Provinces and Berar.

[5] The disappearance of plague in England is also a little mysterious and occurred two centuries before sanitary improvements took place (see Wrigley and Schofield, 1981); changes in methods of house construction, which discouraged rats, may have contributed. For a summary account of plague in India see R.H.Cassen, 1978, p.81-83. It would be wrong to conclude that the exposure to plague was not socially determined even if the dynamics of the disease seem to be partly exogenous; Klein gives evidence of its social differentiation in Bombay (Klein, 1986).

[6] See R. Glass (1948) for an illustration of the tardy implementation of sanitary reform and the high level of infant mortality in the worst areas of an industrial town in the 1920s and 1930s. Several studies, however, have sought to explain the high level of aggregate infant mortality sustained

until the beginning of the twentieth century even after sanitary reform had taken place (see Beaver, 1973; Woods, Woodward and Waterson, 1989).

[7] Similarly, a study undertaken by the author showed that adult mortality in Calcutta in the 1960s and 1970s was lower than in most of the urban districts of West Bengal (Crook and Malaker, 1989).

[8] See for example the first Bengal Sanitary Report of 1864-65 (GOI, Department of Health, 1865).

[9] For a numerical and diagrammatic illustration of these contrasting cases see Appendix A.

[10] As we point out in Chapter III below, differentials in the excess of female over male mortality will also account for the sex differential in urban populations; this also reflects cultural differences.

[11] In the 1972 Fertility Survey (GOI, Registrar General, 1976), fertility in the last twelve months was recorded by state and for rural and urban areas separately. The total fertility rates calculated from these data show urban fertility to be lower than rural fertility in every state.

[12] See Mills and Becker (1986) for a discussion on the post-Independence period.

Table 2.1
Growth of the Urban-dwelling Population and the Total Population by
Decade (per cent per annum)

	1881-89	1891-1901	1901-11	1911-21	1921-31
Urban	1.4	0.7	0.2	0.9	1.8
Total	1.2	0.2	0.7	0.1	1.0
'Urban' minus					
'Total'	+0.2	+0.5	-0.5	+0.8	+0.8
	(3)	(12)	(14)	(10)	(1)

	1931-41	1941-51	1951-61	1961-71	1971-81	1981-91
Urban	2.8	3.5	3.1	3.2	3.8	3.1
Total	1.3	1.2	1.9	2.2	2.2	2.1
'Urban' minus						
'Total'	+1.5	+2.3	+1.2	+1.0	+1.6	+1.0

Note: Bracketed figures from 1881-1931 indicate number of cities for which an
absolute decline in the population is recorded.
Source: Calculated from Government of India, *Census of India, 1881...1991*.

Table 2.2
Registered Crude Death Rate in Central Provinces and Madhya Pradesh,
Rural and Urban, 1881-1987

	Rural	Urban	Ratio of Urban: Rural Rates
1881	32.0	40.7	1.27
1891	35.7	34.4	0.96
1901	23.0	28.5	1.24
1911	33.8	44.9	1.33
1921	43.5	48.7	1.12
1931	35.6	34.3	0.96
1941	32.4	29.1	0.90
....			
1971-73	18.1	10.9	0.60
1981-83	16.7	9.2	0.55
1987	14.6	8.0	0.55

Source: Personal communication from Tim Dyson, and Government of India,
Registrar General, Sample Registration System, *Bulletin*.

Table 2.3
Sex Ratio of Current Migrants to Urban Areas in Selected States, 1981

North India		South India	
Bihar	1.6	Andhra Pradesh	1.1
Gujarat	1.1	Karnataka	1.2
Madhya Pradesh	1.4	Kerala	0.9
Punjab	1.2	Tamil Nadu	1.0
Rajasthan	1.2		
Uttar Pradesh	1.4		
Orissa	1.4		
West Bengal	1.2		
Maharashtra	1.2		

Note: Sex ratios are male-to-female.
Source: Calculated from Government of India, *Census of India, 1981*, Migration Tables (Table D3).

Table 2.4
Sex Ratio of Population in a Selection of Cities, 1911

North India		South India	
Calcutta	2.01	Madras	1.06
Bombay	1.89	Hyderabad	1.04
Allahabad	1.22	Mysore	1.03
Lahore	1.60		

Note: Sex ratios are male-to-female.
Source: Calculated from Government of India, *Census of India, 1911*.

Table 2.5
Literacy of Females by Urban and Rural Areas of Residence

	1901	1931	1961
Cities	5.9	16.2	–
Urban Areas	–	–	37.2
Total	0.7	2.4	13.2
Difference between			
'Urban' and 'Total'	5.2	13.8	24.0
Change over 30 years:			
Urban		10.3	21.0
Total		1.7	10.8

Note: Figures refer to percentage literate of female population aged 15 and over.
Source: Calculated from Government of India, *Census of India, 1901, 1931, 1961*.

Chapter 3

MIGRATION, RECRUITMENT, AND THE INDUSTRIAL STRUCTURE OF CITIES

Introduction

It is a fact not always appreciated, but well-documented, that India's urban growth has never become explosive in any sense of the word. With a natural increase of approximately two per cent annually and an urban growth of less than twice that rate, it follows that the proportion of the total population designated as urban edges up rather slowly from decade to decade (with increases of three to four per cent per decade at current urbanization levels). From this it follows that migration is not immediately the major cause of urbanization (a combination of natural increase and extension of urban boundaries together being the major components). However, these calculations hide enormous local and temporal variation, as well as the dynamics of the process. The essay herein on steel towns and economies of scale probably brings this out most clearly (Chapter 4), while the essay on the historical dimension of urbanization illustrates it further (Chapter 2). Individual cities may grow explosively for a while at increasing rates, the average of which is three or four times the rate of natural increase: but a decade later a lower growth is always attained.

Although current migration is not the major component in growth, it is none the less the prime demographic mover in the urbanization process: for without rural to urban migration or urban boundary changes (and the latter tend to be necessitated by the former) there would be no increase in the proportion of the population classified as urban. Furthermore, the nature of the migration itself, its age and sex composition, has an influence on the demographic structure of individual towns and cities, affecting their natural increase as a result. In the short-run, meaning a decade or less, such influences are pronounced in particular localities, as indeed is migration itself.

In this essay we explore the nature of and reasons for the age and sex characteristics of the migrants. To pick up some of the greatest variation and enable a degree of detailed study, albeit from secondary data, we focus on the rapidly industrialising region of eastern India, selecting districts from West Bengal, Orissa, Bihar and eastern Madhya Pradesh for analysis; we take the period 1951 to 1971 when this industrialization was proceeding most vigorously. However, we will return later to a more comprehensive state-wise analysis in order to bring out the cultural variations that have bearing on the sex ratio of migrants.

The determinants of migration have been studied by many scholars in India (Connell et al., 1976; Bose, 1978; Premi, 1980; Oberai and Singh, 1983) from various different points of view. But the emphasis adopted here is one that has been relatively neglected: an examination of the character of industrialization in its role as a determinant of the sex and age structure of migrants and the distance they travel. By the character of industrialization we refer to the distinction between heavy and light industry with their differing skill requirements and recruitment strategies. This focus again helps towards an understanding of the dynamic process. For the single most outstanding change of relevance that has come about in India since Independence has been the growth of state and private capitalism in industry. With this has come a quantum increase in the demand for wage labour, both permanent and on contract, in the urban areas. The high point in this process was reached in the late 1960s. It is arguable that the nature of the process has changed since then and that by the 1980s much of the labour demand has become channelled through various forms of contracting systems, and that increasingly, at the spearhead of industrialization, there has evolved a more capital-intensive operation. We will begin by focussing on the 1960s.

Migration and the Family: Female in Response to Male Migration

Our basic source for migration age and sex structures is the National Sample Survey of 1964 on Internal Migration (GOI, Ministry of Finance, 1972) (henceforth referred to as NSS). For the males we concentrate on the age profile of the rural to urban migrants', as it is clear from the 1961 census that at least two-thirds of male migrants in our region were of this kind. For the females it was our initial assumption that most migration was due to male migration: what required estimation, we believed, was the appropriate lag structure. To obtain this, female migrants of less than one year's duration, were related to male migrants of less than one year's duration, 1-5 years duration, and 6-10 years duration, in a linear regression model (Appendix C.i). The data used were from selected districts in the eastern region where heavy iron and coal based industry was located in the period of the Second Plan. Interestingly, only two-fifths of the squared variation (R^2) in female migration could be explained on a district-wise basis by male migration. Low though this explained variation may be, it should be noted that the NSS questions addressed to female migrants found that only 12 per cent migrated for marriage purposes and 50 per cent to join family members (including husbands presumably), so that altogether at most only 62 per cent of female migration could be linked directly with male migration. As a check on this a similar regression was done for all the districts of Maharashtra with similar results. As these are not random samples questions of significance do not strictly speaking arise: the co-efficient for male migration of 6-10 years before was very small, and those for the other durations were 0.32 (under one year before) and 0.17 (1-5 years before) (Appendix C.i). This implies that an increase of 1000 male migrants to a town would elicit a response of 320 female migrants in the same year, and a further 170 over the next five years. The same order of magnitude holds for the Maharashtran districts also. We do not have estimates of different degrees of 'marriage response' (as we shall call it) according to the occupation or class of migrant: in the course of this study we follow through the argument that factory-employed migrants (especially those in heavy industry), who are typically from small-farm owning backgrounds, bring their wives into the town after a short time period once their position is secured (or in a minority of cases remain as a split family), whereas other migrant workers, especially those in the non-wage sector, who are more likely to be landless, migrate with their wives in the first instance. This pattern, however, is subject to some variation according to where the migrants come from, and the living conditions in the city to which they go.[1] It follows that the sex ratio among migrants will, over the short run of five years, be variable, dependent upon the predominant characteristics of the migrant streams: to some extent one can predict this ratio from the industrial composition of different urban areas, though clearly the condition in the rural area of origin is an important determinant also.[2] The differences in the sex ratio that persist in the long-run among migrants are of more demographic significance in the eventual population composition of the town, once some degree of stability is achieved.

We have devoted some time to explaining the variance in female migration in so far as it is related to male migration. For all this there will be a proportion of women who migrate alone for jobs or further education, as revealed in the NSS (where 15 per cent were so categorized) and in the 1981 census questions. We discuss the determination of this 'autonomous' (i.e. not 'marriage response') component later.

In so far as female migration is a response, its age distribution is related to the past male migration history: i.e. the NSS female age profiles have to be understood in the light of a combination of the male migration streams that preceded, and those that took place during, the NSS reference period of one year. We also have to remember there is an age differential at marriage (females being younger by about five years in 1961). Hence the age distribution of female migrants is rather older than it would be if it represented a 'response' only to the current migration of male migrants. With these points in mind we offer an interpretation of the age-sex profiles given by the NSS data (Table 3.1). We will refer to migration from within the same district as being of short distance, and that from another state as long distance, though this may not always in fact be the case. Looking first at the rural-urban movements we find the percentages in the 5-17 age groups are roughly similar for each sex in the short distance category, but higher female (than male) percentages for the longer distance migrants. This illustrates 'marriage response' migration, females being younger than males at marriage; their response is to male migration in the 18-24 age-group in the longer distance categories (recorded in the NSS male distribution), together with those joining males who migrated a few years earlier (who are not recorded therefore in the NSS male distribution). Male 'turnover' migration (of very short duration and typically quite local) will be concentrated in the early years of adult age, hence the particularly high proportion of males aged 5-17 in the short distance category compared to the long distance category where the peak age is 18-24 for men. Proceeding from here to the tail end of the NSS age profiles the migration of the 45+ group is quantitatively insignificant, (but not an insignificant proportion of of the total population of this age (migrant plus non-migrant)): the female percentages are higher than the male, and many are undoubtedly widows, migrating to the towns in desperation (Premi, 1980). In Figure 3.1a we have presented the NSS data standardized by ten-year age groups. The preponderance of male migrants in the 15-24 age group stands out immediately. Whereas there is not much difference by sex in the absolute numbers of very young children, nor in the old people. Migrants from urban to urban areas (whose age distribution is illustrated in Figure 3.1b) are more likely to be moving as a family, and the preponderance of adult males is much reduced. Transfers of personnel between departments and firms account for some of this movement no doubt, with older and more senior men and women equally likely to be transferred in this way.

Age and Education in the Structure of the Industrial Workforce

In the economics of production, age is only significant in respect of the command it implies over different technologies. If an economy is expanding rapidly into new industries, or adopting radically new technology in the old ones, and if this implies a need for new skills and new learning for which a secondary education is a sine qua non, then only the younger age groups in an economy are employable. It is a paradox that, in this sense, the stock of labour in India is too old, not too young, since it has a disproportionate number of people over fifteen years of age who are illiterate (and a disproportionate number over 25 who are without a secondary education). The crude literacy rate for Indian men was only 34 per cent in 1961, and not much better at 47 per cent by 1981. In some of the states where the industrial demand for skilled labour was most, the levels of male literacy were lowest, for example Bihar with 31 per cent literate in 1971, Madhya Pradesh with 33 per cent, and Orissa with 38 per cent. The older cohorts will be more costly to train than the younger cohorts.

It is thus conceivable that a modern industrial estate or new town will get a younger age structure than the national because the industrialists actually seek out younger recruits. The neo-classical economist with an eye on static comparative advantage will argue that this kind of industrialization represents an irrational development strategy. He will argue that if basic educational skills are so scarce then industrial specialization should not proceed in those industries that are skill-intensive. This argument fails to address the question of the next generation, and fails to provide guidance on the question of how natural resource endowments change over time, resulting in a changing comparative advantage. Indeed it relegates all education to the rank of a consumption good.

It is possible, though not easy, to see how far particular industries select young labour forces deliberately, since the age structure of migrants by occupation is recorded in special tables of the 1961 census: these cover the cities only (urban areas with over 100,000 population). We have selected about 20 cases here comprising virtually all those in the eastern industrial region focused upon here with supplementary cases from elsewhere: it is impossible not to be selective as the census itself is somewhat idiosyncratic in its coverage of these data for the cities (Table 3.2).

Interpretation of the data must proceed warily. Older industries, if they fail to expand at the rate of the supply of the labour force, will develop an age structure older than the average. The only way to check this is to examine the data in the special migration tables for a select number of cities, where duration of residence is cross-classified by age in particular occupations, and by levels of education; we do this for Ahmedabad below. For the purpose of this analysis we need to exclude those occupations whose age structure is dependent on promotion through seniority; the professional and managerial personnel are likely to be both older.and more educated. In fact we will restrict ourselves to concentrating on the manual employee groups, classified as production workers, workers in transport and communications, and service workers (according to the National Occupation Classification). From Table 3.2 it can be seen that migrant production workers (column 3) in most of the cities are more heavily represented in the younger age group (i.e. 15-34) than are migrant workers in general (column 2). This generally remains true even when we try to take into account the recency of migration to the city (column 1). For example, in the cases of Ranchi, Durg, and Bhopal a substantial proportion of migrants (40-50 per cent) arrived within the three years preceeding 1961 (column 1); yet in all three cases the proportion of production workers in the younger age group is still larger than that in the workforce generally (columns 3 and 2 respectively). The evidence is broadly consistent with our hypothesis that in the new or modernising industries (which are well represented in this selection of cities) recruits are preferred young.

With the caveat in mind of having possibly introduced a bias by not being able to control for recency of recruitment according to occupation, let us look at the occupational age structure for the male migrant production workers in specific occupations in our city sample (columns 4-17, Table 3.2). We will select for illustration the tool-makers and other metal-workers (columns 8-9 and 12-13) who predominate in the heavy industrial region in the east, and the textile workers (columns 4-5) who predominate in the west. The average age of the former two occupations is below that of the latter, and the educational level is higher. In part, however, this reflects the relative age of the capital stock in the two industries at this date (1961): industries employing metal workers (like the steel industry) were, on the whole, set up or expanded more recently. The average age of electricians (10-11) is also lower than the average for production workers in virtually all the cities, and their educational level is higher.

It is particularly instructive to look down the columns under particular occupations in Table 3.2. It should be noticed for example that a particularly young age structure is found exceptionally for textile workers (columns 4-5) in Ulhasnagar near Bombay. This post-Independence industrial development incorporates the latest technology. The proportion of secondary school matriculates in that young age group is far higher than in the rest of Maharashtra's textile workforce, and illiteracy is very low. The other newly developing industry in the west of the country was at that date the chemicals industry (columns 14-15): the new industrial areas of Baroda have a highly qualified workforce in that occupation and a very young age structure. High qualification (little illiteracy) and young age structure also characterize the tool-makers (columns 8-9) at Bhopal (the site of the new electrical engineering industry). In a state with a low educational record on the whole, this cannot be simply the result of a supply of youthful migrants, but a feature of demand requirements. We discuss the iron and steel making industry (columns 12-13) in a case-study below; the youngest age structure and lowest illiteracy occur in the new-town areas of Durg and Ulhasnagar. Coimbatore also has a particularly young and educated workforce in the metal and tool-making industries.

It is also worth looking at the exceptions; for these we should look along the rows rather than down the columns. The private sector company steel town of Jamshedpur is clearly a case. Here the high incidence of educational qualifications is particularly notable, yet the age structure is unusually old. The latter is explained by the paucity of migration between 1951 to 1961 to Jamshedpur, in contrast with the new public sector steel towns. The foundation of the industry here dates from the early years of the century. The Tatas carefully recruited a migrant labour force for work in their mills and then trained it on the spot to a high level of skill; subsequent recruitment took place from within the town, but capacity expansion was carried out with minimal net increase in the labour force and a relative increase in the technical and managerial component, both factors leading to an older workforce than elsewhere in the industry (Datta, 1986).

To point up the contrasts, we have included a very traditional occupational group in the analysis: millers, bakers, and brewers (columns 16-17). Here generally between 50 and 60 per cent of the workers are in the 15-34 age group (as against 60 to 80 per cent in the case of electricians). As might be expected illiteracy is high throughout. Of course this also reflects the fact that bakeries have always existed in the towns even before the impact of modern industrialization: contrast their age structure with that of the tool-makers and metal-workers in Coimbatore, for example. However, there will have been a local multiplier effect when new industries or new towns were created, with an increased local demand for manufactured food and drink, and yet the resultant migration has not selected noticeably the younger cohorts. This can be seen strikingly in the new town area of Durg (which includes Bhilainagar steel town), where the proportion of young millers (65 per cent), though above average for the city sample, is still below the proportion of young electricians (86 per cent); a similar picture is observed in the newly expanding city of Thana.

Finally, and for rather different reasons, it is instructive to look at the residual column in this table, that is those labourers 'not elsewhere classified' (columns 18-19); this will probably consist predominantly of casual labour whose mix of occupations defied the census investigators. It is a large category. It tends to have an older age structure but this varies to some extent depending on the rate of growth of the town in question; illiteracy also varies remarkably. In a few exceptional cases it has a notable proportion of matriculates, around five per cent instead of the more usual one per cent or less (data not shown). In the eastern industrial region Asansol and Jamshedpur stand out in this respect as two cities that acquired their steel industries at an earlier date, and hence built up an educated workforce. The Second Five Year Plan expansions at these plants failed to absorb this stock of qualified labour, and hence follows its comparatively high representation in the residual occupational category. This is a story that may be repeated later in the modern industrial towns that grew up during the Second Plan, as our essay on economies of scale tries to illustrate.

It would be better, however, to have information for particular occupations cross-classified by duration of residence within that large industrial category of production workers. For this detail we are unfortunately restricted to the city of Ahmedabad (Table 3.3). Of the six production occupations we have, the two with the highest literacy rates and a higher incidence of secondary education (data not shown here) are the tailors and the makers of tools and precision instruments. These two groups are also characterized by a younger age structure at both durations of residence recorded. The metal-workers come third. All three are skilled occupations for which the ability to read and basic numeracy would be useful, which would not be true of the factory spinners and weavers for example, who predominate in this city; the latter can be seen to have below average proportions in the younger age group. The differentials are not great, but they are all consistent with what we have observed above.

Let us summarize the discussion so far. The overall age distribution of the workforce in a town will clearly be substantially determined in most cases by the duration of its residence. The declining industries, and the tertiary sector (a labour-absorbing sector) will have the older age structures. The new industries will have young structures, though quickly ageing if subsequent expansion is accompanied by substitution of capital for labour. On the greenfield, or greenish, industrial sites, local industrial workforces are likely therefore to be younger in all economic sectors initially, but especially so in a high technology industry. Skill-specific demand is likely tp turn out to be age-specific (given the incidence of increasing illiteracy with age in the Indian population).

We now discuss the evidence regarding the iron and steel industry that can be got from the developments in Durg district (the site of the Bhilai steel plant) by way of an illustrative case-study. The steel plant was established a few years before the collection of the data we review from the 1961 census, on a greenfield site but in the vicinity of an older locality. This will explain why the workers in transport, trades, and communications are older than their industrial workforce colleagues, for there were passenger buses, though not so many, before the steel workers arrived. By confining our observations to a single industry we tend to overcome this problem of interpretation. The iron and steel industry at Bhilai employs both furnacemen and electricians (Table 3.4). The latter have a higher incidence of literacy than the former. The age structure among the migrant electricians is also slightly younger. What is interesting is that this remains true if we consider the illiterate group alone. For they are younger in the case of electricians than furnacemen, and in both these occupations clearly younger than illiterate construction workers who built the plant. This is consistent with the argument that the greatest potential for training (and economic return on such an investment in human capital) can be found in the younger age groups. It is not only a question of having to recruit young men if you need literacy. Furnacemen, for example, are being recruited young so that they are trainable in the operation of the most modern equipment; construction workers on the other hand need little specific knowledge in the construction of the plant. The construction labourers were not taken on as core-sector workers in the operation of the plant they built, a fact that was often resented; they were regarded as unqualified for a number of reasons, among which will have been health status, age and education.[3]

The future implications of this may be important. The young workforce cohorts in modernising industries will tend to remain with those industries over time: their chances of recruitment into another industry elsewhere will diminish as they grow older. If steel or engineering plant becomes outmoded it will cease to recruit new labour, and its labour force will age. If it faces closure its labour force by virtue of its age will be the least mobile, only qualified to enter alternative industries whose demand for labour is not specifically oriented toward the young. However, it might be regarded as unduly pessimistic to forecast total lack of competitiveness to arise within one generation of the arrival of the initial workforce, though such questions have been raised, if only to be dismissed, in regard to the Durgapur steel plant.[4]

In a country like India with a young age structure the coincidence of redundancy and immobility is less likely; if layoffs occur during stagnation, they will be of a less aged and more deployable workforce on the whole. However, if the growth in new recruitment slows, being confined to replacement levels only, there is a natural tendency for any local labour force to age, witness the case of Jamshedpur (see Table 3.2).[5]

The other implication of recruitment of a young labour force is that a large second generation will be born from it. This will be fired with strong aspirations, given the novelty and promise of industrialization. The requirement is for sustained rapid expansion of new industries, failing which the potential frustration will also be strong. We illustrate this point in greater detail in another essay in this volume (Chapter 4).

Sources of Migration and Recruitment

It has been emphasized from the outset that there is no simple model of the industrial migrants' characteristics. These are determined by an interaction between the demand for, and supply of, labour. Perhaps too much academic analysis has concentrated on the supply side, on migrants looking for jobs, rather than on jobs (or jobbers) looking for migrants.[6] Different types of jobs seek different types of migrants, but it is not equally true to say different types of migrants seek different types of jobs. The landless labourer or small farmer would be eager for employment if the wage were right whether in textiles, or engineering industry: but the two industries would select their labour, sometimes quite carefully, each on rather different criteria. These points have been made and the practice documented in some detail by M.D.Morris (1960) and others. To give two examples: the workforce for the Chittaranjan locomotive works set up in the 1950s has been described as 'a large contingent specially recruited from various parts of the country'; only eight per cent of the workforce was from the local district of Burdwan

(Mohsin, 1964). Similarly in the Bihar coal fields there was said to be 'a long tradition of systematic recruitment of individual workers from outside Dhanbad region'; these were mainly from Uttar Pradesh (Rothermund, 1980; and compare with note 3 above).

As Morris pointed out there is a distinction between the advantage to employers of a floating labour force in the textile industry (less likely to be involved in unionisation) and the advantage of a more permanent workforce in iron and steel (where skills acquired on the job were crucial to the smooth functioning of the production process). Of course there are offsetting disadvantages of permanence here: labour's disruptive power becomes acute when one break in the chain causes the whole process to come to a halt. Worse still the stoppage may cause physical damage to the plant, as when furnaces are allowed to cool too rapidly. The effect of a stoppage by the blast-furnace crane operators at Jamshedpur (a small group of specialized, but technically unsophisticated, workers) is documented by J.L.Keenan (1943), a former manager: substitute crane operators from elsewhere on the site were recruited and productivity at the whole plant fell, it is claimed, about five-fold.

One response to this problem is to reduce the workforce per unit of capital. That is to say, the true cost of labour, including these risks of disruption, may be sufficiently high in some industries to be an inducement to capital substitution: the argument is familiar to specialists in the field of Indian agriculture (Byres, 1981). An alternative response is to recruit a workforce that is more likely to be 'tame'. Recruitment of labour from afar (and preferably from rural areas) may suit this purpose: hopefully the local unemployed, the 'sons of the soil', will fight it out with the migrants in the streets, not on the shop floor. If economies of scale exist at the plant level, and if large plant labour forces are especially risky, then the inducement to substitute is strengthened (Prais, 1983). J. Harriss in his study of Coimbatore notes how 'the most commonly stated reason for starting new small-scale units rather than building up one unit is, broadly, this factor of labour control' (Harriss, 1984, p.191). The increasing ancillarization of Indian industry in the 1980s, the farming out of engineering jobs to small-scale factories for example, was partly prompted by the desire to avoid the accumulation of large plant-based labour forces (Harriss, 1989). Written documentation of such a conscious policy is, maybe for obvious reasons, rather hard to find. Selective reading between the lines is risky, but part of the hue and cry on the conflict between locals and migrants, the 'sons of the soil' issue, can surely be seen in that light. Management clearly resented being told whom to *notify of jobs* despite not being told whom to *recruit*. Witness the chairman of the Indian Iron and Steel Company (who own the Burnpur works in West Bengal) deploring the new legislation regarding local employment exchanges, on the grounds that it was 'fostering the evils of provincialism' instead of promoting the 'mobility of industrial labour...of benefit both to workers and to their employers'.[7] But quite explicit evidence comes from the early days of the Tata Iron and Steel Company's enterprise at Jamshedpur: a biographer of J.N.Tata wrote that he was very keen to 'draw workers from many regions to make the threat of combination and dear wages less likely in the future' (quoted in Datta, 1986). Ironically he was wrong. In the 1920s Sikhs and Biharis joined hands versus the management, and the serious disturbances that resulted may have hastened the building of the civil Township that followed soon after.

Whatever the motives or compulsions facing management, the fact is that rapidly growing heavy industrial towns of eastern India have always recruited labour from outside the district. In the early days TISCO used new local firms as contractors to recruit labour directly from Mayerbanj and Sambalpur districts in Orissa, as well as from Manbhum and the local Singbhum districts in Bihar, thus establishing the precedent of the multilingual steel town in India (S.K.Sen, 1975). Forty years later the steel plant at Rourkela in Orissa was to take labour from Bihar, and also from the south. Bihari labour was also employed by the enterprises at Durgapur, Burnpur, and Chittaranjan in West Bengal. Bhilai in Madhya Pradesh acquired labour from Andhra. These practices of inter-state migration become fuel to the fires of the 'sons of the soil' (Weiner, 1978).

Naturally shortages of educated or skilled labour may require searching far afield. This was more clearly the case in the early days of Jamshedpur, where special construction skills needed in the erection of the furnaces were obtained from among the ship-builders of Bombay Presidency. In addition, white-collar workers came from Bengal, Bombay (mainly Parsees), and Madras. Skilled and semi-skilled labour needed for the production process came from the United States, Germany, and South Wales. This pattern was to repeat itself with the new generation of steel plants built post-Independence, scarce skills being provided by the Russian workforce for Bhilai, West Germans for Rourkela, and so on. Unfortunately the census data do not allow detailed investigation of skills in particular occupations according to the origin of migrants. But we can make some headway by restricting the analysis to the large cities. Jamshedpur is particularly well documented. The integrated steel plant had been expanded prior to 1960. The stock of migrants as of 1961 represents a history of labour migration (18 per cent claiming to have arrived during the period of expansion). Of all the manufacturing workers, 85 per cent are migrant, and 48 per cent came from beyond the boundaries of Bihar. It is difficult to believe there was still a shortage of skilled labour since the city had come to acquire its own skills since the establishment of a captive training institute in 1921. The long-distance migrants are particularly concentrated in the iron and steel manufacturing, coming mainly from West Bengal, Uttar Pradesh, Punjab, and Orissa.

We cannot analyze the other steel towns in quite the same detail. But working mainly from the urban district-level figures we can say that at least two-thirds of the manufacturing labour force was drawn from outside the locality, and over 40 per cent from another state (Table 3.5). Furthermore approximately the same percentage of long-distance migration holds for the non-manufacturing sectors also. It seems therefore that although the desire and need to reach further afield are promoted by the onset of industrialization, especially heavy industrialization, in the process a source of labour supply is mobilized for all workforce sectors.[8] In the case of Durg district (home of the Bhilai steel plant) 58 per cent of the male population comes from another state; rural Andhra and rural Maharashtra are the main areas involved. A similar picture holds for all the steel towns; by way of contrast the more mixed manufacturing, commercial or administrative towns in the heavy industrial region (like Burdwan City) are less likely to be populated by migrants from other states. There is less long-distance migration to the lighter and less skill-intensive industries: the textile towns of western India illustrate this clearly. It seems that large and heavy industrial complexes promote the emergence of a multi-regional urban population, often drawn from a considerable distance.[9] The desire to create a loyalty to the firm by separating the labour from its hearth and home was surely part of the explanation. Labour forces in dispute are far more tenacious if they can readily obtain supplies and support from their local communities closeby. A final anecdote illustrates the importance of labour demand in the distance which the migrants travel: the West German consultants who supervised the construction of Rourkela steel plant claimed there was a lack of carpenters in India and hence imported the necessary workmen from Germany!

Building up a skilled labour force locally may have its own longer term problems. By 1971 the sons and brothers of the migrants were themselves beginning to matriculate. Ranchi, to which we referred above, still stands out in this respect. The proportion of production workers who were matriculate had quadrupled. The proportion of service sector workers so qualified had doubled. And the proportion of 'residual' workers so qualified had increased also, faster than elsewhere in the state. Weiner (1978) wrote of potential social problems such a phenomenon would bring in Ranchi City. Similar observations were made to me by a Chief Mechanical Engineer of Chittaranjan Locomotive Works. Without sustained local industrialization in an atmosphere such as this the potential migrant sees his prospect of employment diminish. And the older migrant sees a much reduced likelihood of his sons following in their father's footsteps at the spearhead of the industrial labour force. At the same time, the prospects of migrating out of these cities, with a special skill or good education much sought after, from somewhere like Bihar to Gujarat, are as remote as are the days when the Gujarati shipbuilders built the furnaces of Jamshedpur.

Sex Composition in the Migrant Workforce

At the start of this essay we attributed most female migration in India to the previous or simultaneous migration of men. The minority of female migration that we referred to as 'autonomous' was left out of the discussion. To these migrants our discussion now returns. We begin by taking an illustrative case-study from the region of the new steel town development at Bhilai. As with the other characteristics of the labour force, the employment of women is not a simple case of a specific industrial demand. A significant feature will be that women are absorbed in the non-factory sector because they are already in the urban locality and will be seeking an addition to household income.

We focus on contrasts between Durg and Bilaspur districts of Madhya Pradesh. Both were subject to high rates of in-migration from 1951-1961; but the former depends on the steel industry while the latter is much more diversified. In the case of Durg, men preceded the women. The demand for labour could, a priori, be regarded as male-biased; there was a tendency to recruit men for jobs such as furnacemen. We note from Table 3.6 that the sex ratio (male-to-female) is 43.5 in the metal goods production sector, which is extremely high. By contrast the managers of tobacco or textile manufacture are keener on selecting women specifically: because the level of training is low, instability in the labour force (as a result of childbearing for instance) can be tolerated, and this in turn facilitates the payment of lower wages. More importantly, with the possibility of the putting-out system, manufacturing work is not incompatible with household work, and this facilitates piecework and even lower remuneration. We observe that the sex ratio in textiles is around 6.2, and nearly evenly balanced (0.9) in the household industry at Bilaspur. In tobacco manufacturing the making of bidis is mainly household based and the sex ratio is 0.6 or 0.7, i.e. three women to every two men; in the non-household sector there are relatively more men. Most of the female household labour in Bilaspur is working in the textile and tobacco industries, and only a minority will be hired-in. The relatively large number of women employed in the lighter industrial town of Bilaspur may reflect an element of deliberate recruitment of women here from outside the town; but it is more likely the case that there is greater use made of women as they are already in the town, having accompanied their migrating husbands in the expectation (and need) of employment. Neither the same expectation nor need would exist among those recruited to work in the heavy industries (with their better remuneration) in Durg. Hence in the household sector of manufacturing the ratio of females to males is higher *in nearly every branch of manufacturing* in Bilaspur than in Durg. It seems as though once the pool of female labour is present manufacturers are able to exercise their preference for cheap labour when it would not be worth their while to go out and recruit directly.

This analysis suggests that sex ratios in the labour force are not purely the result of sex-specific labour demand. There is reason to expect that both 'marriage-response' and 'autonomous' female migration will have some variation related to economic circumstances or cultural characteristics in the locality of origin. That is to say, certain features on the labour supply side will be apparent in the migration characteristics. One feature here would be the greater independence and mobility of women from certain parts of the country. In particular the southern states of India are distinguished well in an analysis of the sex ratio of migrants (Table 3.7). Using the data from the 1981 census we can see how the migration streams into urban areas of the southern states have male to female sex ratios of less than 1.2. All the northern states with the exception of Gujarat and Punjab have sex ratios of more than 1.2. Maharashtra lies on the borderline of this cultural divide and its sex ratio is correspondingly 1.17. West Bengal, also with a ratio of 1.17, is often regarded as an exception to the north-south divide (Dyson and Moore, 1983); the rather low figure for Gujarat is less easy to explain.

A criticism of this analysis might point to the fact that the northern states contain some of the heavier manufacturing complexes where the demand for female labour is likely to be less. Let us look then at the sex composition within specific occupational groups living in urban areas in selected states (Table 3.8). Starting with the male-dominated occupations, we see that blacksmiths and machine-tool operators have an extreme variation in sex ratio across the states. In Bihar, West Bengal, and Madhya Pradesh these lie between about 100 and 150 men to every woman so employed. In the southern states they lie between 20 and 80 approximately. In Tamil Nadu there are 68, and in Karnataka 26. Metal-processors have less variation and the north-south divide is not so clear. The chemical industry workers and the electricians show the expected north-south variation. Turning to the occupations most associated with light industry, we find the north-south divide preserved even here: in Bihar and West Bengal there are over 20 male spinners and weavers to every female, and in Madhya Pradesh thirteen. In Karnataka and Tamil Nadu there are under five. Food-processing also shows the same variation. In general it is clear that social emancipation plays an important role in the urban migration of women, and interacts with patterns of industrial demand.[10]

It must be pointed out that the sex ratio in a population subject to little migration is strongly determined by sex differentials in mortality, which is widely regarded as a measure of women's status relative to men. The sex ratio of the urban migrants, most of whom are internal to the state, will certainly reflect this fact as well as the relative mobility of women.[11]

Another way of analysing this phenomenon is to take individual cities and observe the sex ratio of migrant streams from different states; (unfortunately the data do not allow us to relate these simultaneously to individual occupations). The metropolitan cities of Bombay and Calcutta have a higher than average proportion of single-member, predominantly male, households, reflecting the lack of living space in those cities. The old textile cities of Bombay and Ahmedabad still house their industrial labour in male-only chawls. In these cities the sex ratio of migrants from other states was remarkably skewed in 1961. This is despite the fact that these are not the fastest growing cities, where men often arrive first and form or reconstitute their families a little later. Indeed the situation had changed little by 1981. If we look at the more specific areas of origin of these migrants we find that those from Uttar Pradesh are particularly prone to remain single (Table 3.9a). By way of contrast, migrants to Madras city in the south have a far greater tendency to come from neighbouring states, and to consist accordingly of a much higher proportion of women (Table 3.9b). This is for two reasons; first because local migration streams always contain a substantial number of family migrants (compare with Bombay on this point), and second because the southern states send out more women among their migrants.

The foregoing empirical discussion illustrates the more important factors that determine sex ratios in the labour force and ultimately in the city also. Returning to the state-wise analysis we find there is a wide range of sex ratios among the urban-based production workers (data not tabled): in 1981 the range goes from 73 men to every woman in Punjab, with Uttar Pradesh next at 50, down to seven men to every woman in Manipur, Kerala, and Karnataka. In the industrial states of the eastern region the distribution goes from 31 in West Bengal, followed by 24 in Bihar, to ten in Orissa and Madhya Pradesh. No doubt each state's sex ratio could be broken down into its component parts, distinguishing the contribution made by composition of industry, recruitment and housing policy, and geographical or cultural origins of the migrants; but we do not attempt this major exercise here.

The state-wise urban sex ratios reflect the foregoing and other factors. The range goes from 1.4 in Sikkim to 1.0 in Kerala. Altogether these factors comprise the following: (i) differences in fertility, since higher fertility results in higher proportions of local births and hence of non-migrant population, which means, ceteris paribus, more even sex ratios; (ii) differences in mortality, since in India higher overall mortality will usually coincide with higher female than male mortality, the outcome being fewer surviving females; (iii) differences in current migration levels and characteristics, which are the net effect of in-migration (on which we have written a lot here) and out-migration; (iv) and maybe an interaction between these factors. This would be a rather complex system to model. Here we attempt something simple to explain the urban sex ratios by state. As explanatory variables we take the overall sex ratio in the same state, which represents differential mortality and cultural effects on local migration; the second variable is the proportion of large (class I) city populations in the urban population of the state, which is intended to represent the influence of heavy industry and housing constraints; and the third variable is the urban growth rate, which picks up differences in natural increase and in net migration into the state (see Appendix C.ii). Only the first two variables seem to be systematically contributing to a comprehensive explanation. Changes in the overall sex ratio of the state give rise to more than equal changes in the urban sex ratio—not surprisingly in so far as this variable represents cultural factors that determine the sex ratio of migrants. A much smaller change results from a one per cent increase or decrease in the proportion of the large cities, and the direction of the effect is unexpected: larger cities lower the male-to-female ratio; and in other models that we tried out the significance of this variable was lost. Altogether about three-quarters of the variation in urban sex ratios is accounted for in this model, The strongest effect comes from those characteristics of the state that affect its overall sex ratio—for instance differential mortality. We are not able to detect much effect of the characteristics of industrialization at this level of aggregation. In a further model we used the sex ratio among current migrants as an explanatory variable: this could reflect both the characteristics of labour demand and those of labour supply and hence interpretation is ambiguous (Appendix C.iii). This variable proves to be significant, but its effect is not as quantitatively important as that of the sex ratio of the state, which underlines our contention that, at the aggregate level, urban growth is not predominantly the immediate result of migration, but of natural increase.

Participation of the Low-caste Groups in the New Industrial Ventures

It is unfortunately not possible from the available materials to analyze the class composition of the different types of urban area, nor to document how these areas might be changing in respect of class structure over time. Most of our analysis has had to be restricted to the demography of the aggregate urban population, with some extra detail in the case of migrant populations only; and most of this has had to relate to locational rather than economic background of the migrant, though some attempt has at times been made to deduce the latter from the former. However, some limited demographic information is available to us from the census regarding the scheduled castes and scheduled tribes (which we shall call collectively the lower castes); as these are usually the lowest socio-economic group we can at least take their recorded experience as an indicator of the interplay between demographic and certain social class factors.

The lower castes participated considerably less than the rest of the population in the industrial expansion of the heavy industrial areas developed in the Second Plan period (Table 3.10). This situation continued over time. For example, by 1971 there were still consider ably fewer lower caste workers reported as employed in factory-sector industry in the Bhilainagar-Durg urban agglomeration than were recorded from the general population (46.2 per cent compared with 61.5 per cent).[12] It is to be noted that in the district of Durg the participation of this social group in manufacturing is mainly in towns like Rajnandgaon and Dongargarh, where prosperity was least sustained, to judge from their growth rates.

The situation is similar but worse in the steel town of Durgapur. Only 37.6 per cent of lower caste workers were participating in factory-sector manufacturing in 1971, compared with 56.3 per cent of the general population. We should note how their presence in the city increased over the decade from six to twelve per cent of the population (if the Census records are reliable in this difficult area of investigation), but their participation in the main industrial sectors remained substantially less than that of the general population. By 1981 the scheduled castes and tribes still only made up 13 per cent of the city's population, despite being 25 per cent of Burdwan district. This enhances the effect of an emerging duality in these new cities. The evidence is consistent with that from a survey conducted by the author in Durgapur in 1988. Landless labourers, in contrast with the landed, had mainly migrated to Durgapur after the inauguration of the steel plant, and were more likely to be employed in trades and services than as core operatives at the plant. For the urban District as a whole the gap actually widened between the lower and higher caste groups in terms of participation in factory manufacturing.

The picture from outside the heavy industrial area is not dissimilar. In the new-town localities of Maharashtra there are also low proportions of the lowest castes; the percentage in Pimpri-Chinchwad, a modern and rapid-growth mixed industrial development outside Poona, was only half that in the older and more slowly growing towns like Sholapur. A study within Poona itself, where the proportion of low caste groups was about seven per cent, showed that in some slum communities up to 50 per cent of the population was from these groups; these were typically the communities that had been forced out of the drought-prone districts of the state (Bapat, 1981). These migrants come as complete families from districts not far inland, and once located in the city are effectively locked in as they have no rural land. If the city fails to prosper they are the least likely to return to a rural livelihood, and the least able to migrate to another city. This most depressed group is characterized by a high female to male sex ratio and a mature family structure from the start. Their women are more likely to be employed than is the case in the general population, partly because the lowest castes are more likely to participate in branches of manufacturing that are less likely to select men alone (Table 3.11); there are fewer to be found in steel production for instance, which may explain why the sex ratio in the overall workforce is as high as 14.9 (men-to-women) in manufacturing in Durg, but only 2.8 among the lowest castes. But, more importantly, it is the case that women from the low caste and usually poorer households are forced through economic need to work outside the home when other women would not have to. Note from Table 3.11 the contrasting sex ratios in the construction industry in Durg, with nearly four men (3.7) to one woman from the general population, but less than two to one (1.5) from the scheduled tribes and scheduled castes.

The lowest caste groups predominate in the mining areas (Tables 3.10, 3.11). And it is they who remain when the mines close down or fail to expand. In mining, as we shall see in another essay, there is a tendency for migrants to achieve some sort of occupational mobility with time. The older the mine the more local its workforce and the greater the participation of the lowest caste groups. So there becomes a concentration of a class of people who are already the most vulnerable in the rural areas, being typically landless. Equally, the older the mine the more likely it is to be closed - the equivalent disaster to a harvest failure, but taking place in an urban monoculture with a clearly differentiated class of wage labourers. At the community level the problem is exacerbated by the older age structure of a declining population. It seems that the scheduled castes and tribes stood to gain least from the worst consequences of rapid industrialization.

The failure of the lowest caste groups to be represented in the developing localities can be observed to be long-standing. In the city of Jamshedpur, founded in the first decade of the century, only 13 per cent of the population was recorded as scheduled caste or tribe as late as 1971; yet in the local district of Singbhum 62 per cent of the rural population are scheduled castes and tribes. This is an extreme case. But the observation strengthens the argument that recruitment managed, consciously or unwittingly, to miss the local labour. Those that did get to the city, however, found employment in manufacturing in almost the same proportions as the general population. We can never prove the case as being one of supply rigidities or demand rigidities;[13] what remains striking, however, is the small number of tribals in this long-standing industrial city in the heart of a tribal area.

We have presented some contrasting data for other localities in Bihar (Table 3.12). The failure to employ tribals in the developing areas in preference to migrants from outside (mainly West Bengal) was a highly contentious issue in the 1960s. Despite the high proportion of tribals in Ranchi and Singbhum districts their urban representation is small. In Bokaro steel town area local Bauri tribals were employed in compensation for their loss of land, but only as temporary unskilled workers: the terms of contractual employment with Hindustan Steel required literacy for many occupations (though the Census records illiterates in what appears to be the core-sector workforce), thus effectively disqualifying the tribals. This underlines the point that, whatever policies may be instituted to promote the employment of the scheduled castes and tribes, little will be achieved unless they have equal access to basic needs when young: particularly health care and schooling. In Dhanbad urban agglomeration the proportion of tribals represented is only half that in the district, and those are predominantly in mining. The picture is one familiar elsewhere in India: the lowest castes are under-represented because the occupations in which they find them selves locked are under represented (D'Souza, 1975). They are to be found in the mining and declining towns, where the more socially mobile have moved out.

We have described earlier how India's industrialization underwent a structural change in the 1960s and 70s. It may be argued that this is what is manifest in the contraction and expansion of specific industries. With it has come the expansion and stagnation of towns and cities. The most disadvantaged social classes, though undoubtedly gaining something from the urbanization process, seem to have gained less than the labouring classes generally. By the end of the 1970s a further change had begun to take place, also indicative of changes in the organisation of industrial capital, which was to the disadvantage of labour generally. When in the late 1960s and early 1970s organized labour had put pressure on capital for better working conditions and higher wages, capitalist industry had begun to respond by contracting out various stages in the manufacturing process. This has been well documented for the transport equipment industries. At the same time the continuous process industries like iron and steel making and electricity generation began to resort to, or intensify their use of, contract labour for tasks such as cleaning furnaces and transporting fuel within plant. Pay and conditions of service, as well as the opportunity to mobilize in order to improve them, are at their worst in the case of contract labour. It is reported that the Rourkela steel plant was employing over 8,000 contract labourers in the 1980s (Sengupta, 1985) in addition to its permanent workforce of 40,000. They are recruited disproportionately from among the scheduled tribes of the region and include a disproportionately large number of women.

Through disaggregation of the workforce in this way it is possible to come to a clearer appreciation of the experience of particular social groups. To some extent the figures misrepresent social mobility in so far as there has been ascription into the scheduled caste category of those whose families were not of scheduled caste origin to gain from reservation policies. But from 1961 to 1971 (when ascription was less important than later), while the general urban labour force in factory manufacturing grew at 2.9 per cent per annum, the lower castes only managed 2.1 per cent (with the scheduled tribes at 3.9 per cent doing better than average). Disaggregation has revealed that low caste participation in the core industrial sector has been less than that of the general population, whereas its participation in the older industrial areas and in mines especially has often been above average.

Conclusion

What we hope to have brought out in this essay is that migrants are not an undifferentiated stream of men and women leaving depressed rural areas in the hopeless search for urban work. The role played by industry itself in selecting and encouraging migrants of specific age, sex, and background is, we hope, overwhelmingly apparent from the illustrations we have given. The diversity in the composition of migrants corresponds with the diversity in industrial structure: hence towns that are dominated by a particular industry will end up with a particular demographic structure. Clearly the age and sex structure of migrants (in conjunction with the size and time-phasing of the migration streams themselves) will have short and long-term demographic consequences; in Appendix A to this volume we simulate several scenarios to illustrate how such relationships work out. There will also be resulting mixtures of regional and caste composition in the cities. The consequences are of little concern to the individual industries involved in raising the initial labour force. They are social rather than private. Capitalist industry may gain from the existence of a pool of cheap labour of a particular educational composition, though as we have seen it may fear the potential cohesion of a locally well established population. But the fluctuations in the size of this pool (which follow demographically from earlier decisions), and hence in its price, will not necessarily coincide with phases of investment in the future development of an urban industrial centre. A state that wished to optimize in this respect would need to develop strategies to encourage specific patterns of industrial location through time. Those involved in the shaping of such strategies would need to know well the relationship in all its detail between industrial labour demand, migration, and the demographic composition of cities.

NOTES

[1] This is supported by a survey undertaken in Bombay which indicates that 79 per cent of females who joined other family members (mainly husbands) did so once the latter had secured employment, and only 11 per cent came before a job was secure; whereas in the case of casual workers 60 per cent joined husbands with jobs, and 33 per cent came with them in speculation (Deshpande, 1983).

[2] A survey carried out in Poona throws some light on the economic backgrounds of migrants in different urban industrial employments or occupations. In a slum settlement in which over two-thirds of the household heads surveyed were in factory industry, municipal service, or working as skilled labour, 44 per cent of the surveyed migrants still retained rural land (less than two acres in 50 per cent of these cases). In a contrasting slum settlement where two-thirds of the household heads were in petty trade or casual service, only 12 per cent had land (although 87 per cent had previously been employed in agriculture). A higher proportion of the latter than the former migrants were from local areas (Bapat, 1981). Similar findings come from the Bombay survey referred to above (Deshpande, 1983). Of the permanent factory workers, 72 per cent owned land, with an average holding of 2.7 acres, whereas of the casual workers 51 per cent owned land, with an average holding of 2.1 acres. Surveys conducted by the author and others in the steel-town of Durgapur reinforce this general impression. Of those households whose head was employed by the Steel Authority itself 51 per cent had land in the locality from where they came, whereas of those whose head was employed in the non-factory sector only 36 per cent had land (unpublished data from survey reported in Crook and Malaker, 1988).

[3] One of the public sector steel plant personnel managers is reported to have said that men from Uttar Pradesh were favoured as furnacemen because of their stature.

[4] The same observations can be made from the experience in the western world. On a visit to Corby steel plant in the U.K. in the early 1980s the author found it reported in the local press that steel workers, recently made redundant on the closure of most of the integrated iron and steel works, were unlikely to find employment in the local county of Northampton. There was new industrial growth taking place in this county, but the new industries, light manufacturing and electronics, favoured a young workforce. The sons of the steel workers would be able to partake in such opportunities. By contrast, no such opportunities existed in generally more depressed north-east England when the Consett steel works closed at roughly the same time.

[5] This follows from a demographic principle similar to that of the convergence to a 'stable' population. For even if a specific labour force consists initially of workers in one age group only (say 20-30), the process of mortality, retirement, and recruitment of 20-30-year olds for replacement purposes will gradually result in an older and more stable age structure; it will be older still if recruitment only allows for replacement and no new growth.

[6] My main criticism of literature focusing on the supply side (e.g. Todaro, 1969) is that it ignores rigidities and constraints that were, and still

are, the very essence of the Indian labour market. The probability of getting an industrial job is very slight for a villager with no urban contacts; but very high for a villager whose locality is visited by an agency recruiting for TISCO or SAIL. Hence the urban unemployment rate is likely to be a very weak indicator of job prospects in industry, especially in 'unstable' (i.e. non-steady state) urban populations such as are described in this book.

[7] See *Economic Weekly,* 26.11.60, Speech from the Chairman of IISCO, p. 1732.

[8] This phenomenon may suggest that the relevance of the Todaro migration model to which we referred above is somewhat enhanced the more industrialized an economy becomes; with better knowledge of the opportunities a freer flow of labour takes place; and employers can get the labour they want by drawing on the urban 'reserve'.

[9] There may be some preference for rural labour. There is evidence from western European surveys that it is perceived to be more docile than urban labour (Scherer et al., 1975). What might be more important in the Indian context would be its ability to sustain a strike or lockout (see further our essay here on industrial concentration). The evidence here is rather mixed. The steel authorities claimed to have advertised jobs in urban areas, but the majority of the long-distance migrant workforce is rural in origin, except at Rourkela. It seems from case-study material that the initial core-sector labour force was substantially urban in origin, probably because education was preferred if not insisted upon (Ray-Chaudhuri, 1990).

[10] A more economistic explanation for the phenomenon of more balanced sex ratios among migrants from the south has been suggested by Mark Holmstrom. He speculates that the higher incidence of landlessness promotes family migration rather than single member (typically male) migration that occurs more frequently in the north (Holmstrom, 1976).

[11] In a statistical exercise we found a systematic relationship between the sex ratio of migrants to urban areas and the sex ratio in the population of the state (in most cases predominantly reflecting differential mortality).

[12] In 1960 the Chairman of Bhilai Steel Plant announced his intention to fulfil the nationally stipulated target of filling 15 per cent of jobs at the plant from the scheduled castes and 12.5 per cent from the scheduled tribes, totalling 27.5 per cent of jobs reserved for the lower castes. It is claimed by the Public Relations Department at Bhilai that about 12,000 lower castes have been employed, or 20 per cent of the 1984 workforce (Steel Authority of India Limited, 1984). While less than targetted, this is not a bad proportion considering that the lower castes form 20 per cent of the District's population. However, for reasons mentioned below, we regard the 1971 Census figures as rather more indicative of the degree of social mobility.

[13] Literacy levels among the lower castes have been considerably below average; e.g. the urban male literacy rate was only 32 per cent for the lower castes in 1961 as against 57 per cent for the general population. This clearly restricts the opportunity for employment of the scheduled castes and tribes in industry where literacy is required. The gap narrowed, however, between 1961 and 1981. The difficulty in finding literate workers may be used as an excuse for not employing lower castes, regardless of whether it is true.

Table 3.1
Percentage Age Distribution of Current Migrants from Rural to Urban
Areas, 1964

				Age Groups				
	<1	1-4	5-17	18-24	25-34	35-44	45+	Total
Males								
From same District	3.1	8.2	33.3	18.7	20.0	9.2	7.4	100.0 (2127)
From another District	3.3	7.7	18.6	37.4	18.4	9.8	4.9	100.0 (1134)
From another State	1.7	5.4	17.7	33.3	23.2	11.4	7.3	100.0 (655)
Females								
From same District	4.4	11.6	34.3	20.3	14.4	6.4	8.6	100.0 (1543)
From another District	8.9	9.9	28.9	23.1	16.8	4.5	7.9	100.0 (633)
From another State	5.8	11.3	27.8	23.7	18.2	7.0	6.2	100.0 (302)

Note: Row total in brackets.
Source: Government of India, Ministry of Finance, Department of Economic
Affairs, National Sample Survey, *Tables with Notes on Internal Migration*,
(No. 182, 1964).

Table 3.2
Age Structure and Literacy of Migrants to Selected Cities, in Selected Occupational Groups, 1961

City	1	2	3	4	5	6	7	8	9	10	11	12	13	14	15	16	17	18	19
WEST BENGAL																			
Calcutta	11	55	58	63	27	57	29	61	28	64	31	–	–	–	–	–	–	55	55
Howrah	17	56	59	57	55	72	63	58	41	–	–	66	58	–	–	–	–	61	69
Asansol	13	64	68	–	–	52	27	73	17	76	13	–	–	–	–	49	38	63	50
Burdwan	13	53	56	–	–	58	32	65	32	–	–	–	–	–	–	59	59	51	58
BIHAR																			
Gaya	31	49	57	62	40	61	57	62	30	64	6	–	–	–	–	56	52	54	41
Patna	31	63	60	–	–	52	23	65	15	71	10	–	–	–	–	54	28	63	69
Jamshedpur	18	50	49	–	–	–	–	48	8	45	6	37	13	–	–	–	–	54	46
Ranchi	43	58	60	49	23	–	–	68	13	73	8	–	–	–	–	48	31	60	60
MADHYA PRADESH																			
Durg	51	76	78	–	–	–	–	83	10	86	13	84	26	–	–	65	32	72	70
Bhopal	44	65	69	59	32	–	–	88	6	87	11	–	–	–	–	56	35	64	81
GUJARAT																			
Baroda	27	56	58	53	30	54	11	70	12	–	–	66	38	75	15	–	–	53	47
Ahmedabad	18	52	50	46	39	58	19	62	25	–	–	66	36	–	–	–	–	62	68

Key: (1) % of all male migrants arrived in last 3 years; (2) % of male working migrants aged 15-34; (3) % of male migrant production workers aged 15-34; (4) % of Textile Workers aged 15-34; (5) % of (4) who are illiterate; (6) % of Tailors aged 15-34; (7) % of (6) who are illiterate; (8) % of Tool-makers aged 15-34; (9) % of (8) who are illiterate; (10) % of Electricians aged 15-34; (11) % of (10) who are illiterate; (12) % of Metal Workers aged 15-34; (13) % of (12) who are illiterate; (14) % of chemical workers aged 15-34; (15) % of (14) who are illiterate; (16) % of Millers aged 15-34; (17) % of (16) who are illiterate; (18) % of Other Labourers aged 15-34; (19) % of (18) who are illiterate.

Table 3.2
(Continued)

City	1	2	3	4	5	6	7	8	9	10	11	12	13	14	15	16	17	18	19
MAHARASHTRA																			
Bombay	17	59	61	57	38	62	19	65	19	69	9	59	34	–	–	63	49	61	55
Thana	34	62	66	69	28	63	13	75	11	72	5	68	28	–	–	61	42	62	55
Ulhasnagar	10	59	67	83	20	74	58	75	15	83	3	77	44	–	–	51	31	62	24
Nasik	24	52	53	–	–	51	3	57	16	–	–	–	–	–	–	54	35	46	51
Nagpur	22	50	47	39	39	56	8	51	14	59	10	49	34	–	–	51	31	48	57
Poona	21	50	51	45	38	51	17	61	17	60	7	55	46	–	–	58	28	47	59
Sholapur	21	46	46	45	59	55	19	50	21	54	9	48	40	–	–	42	35	47	56
TAMIL NADU																			
Coimbatore	28	57	59	53	29	61	11	72	10	57	6	71	19	–	–	54	27	55	63
Madras	27	54	53	42	23	60	18	63	11	62	8	59	27	58	23	58	18	51	57
Salem	30	51	49	44	46	55	25	61	21	–	–	–	–	–	–	52	50	51	77
Tiruchchirappalli	28	52	54	46	39	60	17	56	17	52	5	–	–	–	–	56	27	52	60

Key: (1) % of all male migrants arrived in last 3 years; (2) % of male working migrants aged 15-34; (3) % of male migrant production workers aged 15-34; (4) % of Textile Workers aged 15-34; (5) % of (4) who are illiterate; (6) % of Tailors aged 15-34; (7) % of (6) who are illiterate; (8) % of Tool-makers aged 15-34; (9) % of (8) who are illiterate; (10) % of Electricians aged 15-34; (11) % of (10) who are illiterate; (12) % of Metal Workers aged 15-34; (13) % of (12) who are illiterate; (14) % of chemical workers aged 15-34; (15) % of (14) who are illiterate; (16) % of Millers aged 15-34; (17) % of (16) who are illiterate; (18) % of Other Labourers aged 15-34; (19) % of (18) who are illiterate.

Notes:– denotes data not available. Column (2) excludes administrative and professional occupations. Odd numbered columns from (5) to (19) are calculated as follows: illiterate 15-34 year olds in the stated occupation divided by all 15-34 year olds in the same occupation, times 100.

Source: Calculated from Government of India, *Census of India, 1961*, West Bengal, Bihar, Madhya Pradesh, Gujarat, Maharashtra, Madras, Migration Tables (Table D4).

Table 3.3
Age Distribution of Male Migrants to Ahmedabad in Selected
Occupations, 1961

Occupation	Age Group	Duration of Stay	
		One Year	1-5 Years
All production	15-34	76	80
workers	35-59	21	19
Spinners and	15-34	72	79
weavers	35-59	25	20
Tailors	15-34	86	81
	35-59	12	16
Metal workers	15-34	81	82
	35-59	18	16
Tool makers	15-34	84	82
	35-59	15	17
Carpenters	15-34	76	79
	35-59	21	20
Bricklayers	15-34	79	77
	35-59	17	20

Note: Age distributions are percentages but sum to less than 100 because we
have omitted the under-15 and the 60+ age groups.
Source: Calculated from Government of India, *Census of India, 1961*, Gujarat,
Special Migration Tables, Ahmedabad City (Table 4).

Table 3.4
Age Structure and Literacy of Male Migrants to Durg District (Urban)
in Selected Occupations, 1961

	Total	Illiterate	Literate Only
Furnacemen			
Age group			
0-14	0.1	0.0	0.3
15-34	84.0	77.6	82.7
35-59	15.4	21.6	16.5
60+	0.5	0.8	0.5
All ages	100.0	100.0	100.0
	(5090)	(1425)	(1343)
Electricians			
Age group			
0-14	0.1	0.5	0.1
15-34	86.2	79.5	84.0
35-59	13.6	19.5	15.9
60+	0.1	0.5	0.0
All ages	100.0	100.0	100.0
	(3069)	(420)	(923)
Construction workers			
Age group			
0-14	1.1	1.8	0.5
15-34	72.1	69.1	72.4
35-59	25.5	27.4	26.2
60+	1.3	1.7	0.9
All ages	100.0	100.0	100.0
	(2339)	(1203)	(784)

Note: Distributions are in percentages; absolute numbers are bracketed.
Source: Calculated from Government of India, *Census of India, 1961*, Madhya
Pradesh, Migration Tables (Table D4).

Table 3.5
Percentage Distribution of Male Factory Manufacturing Workers in
Steel Town Localities by Areas of Origin, 1961

	Sundargarh (Urban)	Durg Town Group	Jamshedpur	Burdwan Steel Towns	Burdwan City
Born in same locality	33	15	16	27	49
Migrant from same District					
Urban	7	1	2	3	4
Rural	4	12	8	15	9
Migrant from other District of State					
Urban	7	5	3	4	3
Rural	2	6	16	8	8
Migrant from other State					
Urban	38	18	14	9	5
Rural	8	43	40	34	21
Total	100	100	100	100	100
	(18816)	(78267)	(46251)	(48567)	(3855)

Notes: Sundargarh (urban) contains Rourkela, Durg Town Group contains
Bhilai, Burdwan Steel Towns include Durgapur and Burnpur; Burdwan City is
taken as a contrasting non-steel town; for Durg Town Group the data refer to the
male population, not the factory manufacturing workforce alone, owing to the
relevant data not being obtainable; absolute numbers are given in brackets.
Source: Calculated from Government of India, *Census of India, 1961*, Bihar,
Orissa, West Bengal, Migration Tables (Table D6), and Madhya Pradesh,
Migration Tables (Table D5).

Table 3.6

Sex Ratios in Selected Industries in the Urban Areas of Bilaspur and
Durg Districts, Madhya Pradesh, 1961

Industry	Bilaspur		Durg	
	Household	Factory	Household	Factory
All Industries	1.3	3.9	1.7	5.2
	(8994)	(46868)	(5382)	(101592)
Tobacco				
manufacture	0.7	1.7	0.6	0.7
Cotton textile				
manufacture	0.9	–	–	6.2
Leather goods				
manufacture	3.0	3.8	3.6	4.5
Metal goods				
manufacture	4.2	50.0	7.6	43.5
Construction		4.6		3.7

Notes: Sex ratios are male-to-female; bracketed figures are absolute numbers; no
entry is given for small numbers.
Source: Calculated from Government of India, *Census of India, 1961*, Madhya
Pradesh, General Economic Tables (Table B4).

Table 3.7

Sex Ratio of Migrants of less than One Year's Duration of Stay in
Urban Areas of Selected States, 1981

Northern States		Southern States	
Bihar	1.55	Andhra Pradesh	1.11
Gujarat	1.07	Karnataka	1.19
Haryana	1.21	Kerala	0.89
Himachal Pradesh	1.32	Tamil Nadu	0.99
Madhya Pradesh	1.37		
Maharashtra	1.17		
Orissa	1.43		
Punjab	1.17		
Rajasthan	1.24		
Uttar Pradesh	1.37		
West Bengal	1.17		

Notes: Sex ratios are male-to-female.
Source: Calculated from Government of India, *Census of India, 1981*, Migration
Tables (Table D2).

Table 3.8
Sex Ratios of Urban Workers in Selected Occupations in Selected
States, 1981

| Occupations | Northern States | | | |
	Orissa	Bihar	Madhya Pradesh	West Bengal
Metal processors	80	42	15	104
Chemical processors	10	44	27	32
Spinners and weavers	6	24	13	29
Food and beverage manufactures	5	24	11	21
Blacksmiths and tool makers	22	154	98	145
Electricians and electronics workers	352	429	103	57
Bricklayers	8	48	6	27
Transport operators	109	232	84	333

| | Southern States | | | |
	Tamil Nadu	Karnataka	Andhra Pradesh	Kerala
Metal processors	57	79	53	117
Chemical processors	16	12	11	29
Spinners and weavers	3	4	3	2
Food and beverage manufacturers	4	5	6	3
Blacksmiths and tool makers	68	26	47	77
Electrical and electronics workers	47	20	46	33
Bricklayers	6	8	15	12
Transport operators	296	239	190	1070

Note: Sex ratios are male-to-female.
Source: Calculated from Government of India, *Census of India, 1981*, Economic Tables (Table B20).

Table 3.9a
Sex Ratio of Migrants to Bombay of 1-5 Years Duration of Stay, 1961

State of Origin by Birth	Rural	Urban
Maharashtra	1.6	1.2
Gujarat	1.4	1.2
Uttar Pradesh and Bihar	5.0	3.0
Karnataka	1.7	1.3
Tamil Nadu	1.9	1.6
Kerala	3.2	2.2

Note: Sex ratios are male-to-female.
Source: Calculated from Government of India, *Census of India, 1961*, Maharashtra, Special Migration Tables, Greater Bombay (Table 2).

Table 3.9b
Sex Ratio of Migrants to Madras of 1-5 Years Duration of Stay, 1961

State of Origin by Birth	Rural	Urban
Tamil Nadu	1.1	1.1
Andhra Pradesh	1.1	1.1
Kerala	2.1	1.6
Karnataka	1.5	1.1
Other States	1.2	1.0

Note: Sex ratios are male-to-female.
Source: Calculated from Government of India, *Census of India, 1961*, Madras, Special Migration Tables, Madras City (Table 2).

Table 3.10
Participation of Lower Castes in Industrialising Areas of Eastern India

District (Urban) or Locality	Per Cent of Lower Castes in Population		Per Cent of Male Lower Caste Workers who are in Manufacturing/ Mining.
	1961	1971	1971
Durg District	9.8	8.9	29.1 (38.5)
Bhilainagar Steel Town	7.1	6.1	46.2 (61.5)
Rajnandgaon	6.7	8.3	45.9 (29.0)
Dongargarh	7.1	12.2	21.3 (12.3)
Chirmiri Mining Colony	20.1	28.8	83.1 (76.9) (mining)
Surguja District (mining)	17.1	18.5	49.8 (37.3) (mining)
Burdwan District	13.3	11.1	29.3 (35.8)
Durgapur Steel Town	5.8	12.3	37.6 (56.3)
Chittaranjan (locomotives)	7.0	6.4	74.5 (85.1)
Jamuria (mining)	22.5	34.5	22.1 (28.2) (mining)
Raniganj	35.1	29.1	40.3 (38.5) (mining)
Burdwan City	10.6	7.8	15.5 (11.4)

Notes: Figures in brackets indicate percentages for the total population. Lower Castes are the addition of Scheduled Castes and Scheduled Tribes according to the census returns for 1961 and 1971.
Sources: Calculated from Government of India, *Census of India 1961*, Madhya Pradesh, and West Bengal, General Population Tables (State Primary Census Abstract), and *Census of India, 1971*, General Population Tables (State Primary Census Abstract), and Social and Cultural Tables (Table C8).

Table 3.11
Industrial Participation of Lower Castes in Three Contrasting Districts
of Madhya Pradesh, 1961

District (Urban)	Sex Ratio of Workers	Per Cent of Male Lower Caste Workers who are in Factory Manufacturing	Sex Ratio in Factory Manufacturing	Per Cent of Male Workers in Mining/ Construction	Sex Ratio in Mining/ Construction
Durg (heavy industry)	1.9 (4.2)	19.5 (37.9)	2.8 (14.9)	12.4 (10.0) (Construction)	1.5 (3.7)
Bilaspur (light industry)	1.5 (2.6)	9.4 (9.9)	1.9 (4.6)	–	–
Surguja (mining)	3.8 (7.0)	2.1 (4.7)	4.1 (12.8)	45.5 (30.0) (mining)	20.4 (23.3)

Notes: Figures given in brackets refer to the total population. All sex ratios are male-to-female.
Source: Calculated from Government of India, *Census of India 1961*, Social and Cultural Tables (Table C8).

Table 3.12
Lower Castes in Contrasting Cities of Bihar, 1971

City	Per Cent of Lower Castes in City Population	Per Cent of Lower Castes in Rural Population
Established Cities:		
Monghyr	6.5	17.9
Jamalpur	10.6	17.9
Gaya	8.6	25.8
Bhagalpur	6.6	15.3
Jamshedpur	12.8	61.6
Developing Cities:		
Bokaro	22.2	32.3
Dhanbad	16.4	32.3
Ranchi	17.3	69.3

Notes: The rural population is that of the District in which the city is situated. The lower caste population is the addition of the Scheduled Castes and Scheduled Tribes as recorded in the Census.
Source: Calculated from Government of India, *Census of India, 1971*, Bihar, General Population Tables (State Primary Census Abstract).

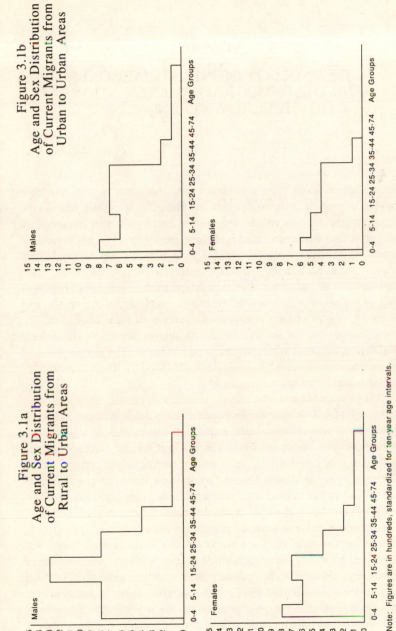

Figure 3.1a
Age and Sex Distribution
of Current Migrants from
Rural to Urban Areas

Figure 3.1b
Age and Sex Distribution
of Current Migrants from
Urban to Urban Areas

Males

Females

Males

Females

0-4 5-14 15-24 25-34 35-44 45-74   Age Groups

0-4 5-14 15-24 25-34 35-44 45-74   Age Groups

Note: Figures are in hundreds, standardized for ten-year age intervals.

Source: Data are from Government of India, Ministry of Finance, Department of Economic Affairs, National Sample Survey, *Tables with Notes on Internal Migration*, (No. 182, 1964).

Chapter 4

TECHNOLOGY OF INDUSTRIAL SCALE AND THE DEMOGRAPHIC IMPLICATIONS: THE STEEL TOWN PHENOMENON

Introduction

Demographers have been concerned with the phenomenon of stable population growth at the national level. This, if sufficiently fast (so the argument goes), gives rise to a supply of children whose demands on the social infrastructure could become prohibitively costly, and of labour which may become surplus to a nation's need (Coale and Hoover, 1958; McNicoll, 1984). Of equal importance, however, is the phenomenon of the non-stable population at the local level resulting from the periodic inflows of migrants, who create, directly and indirectly, waves of labour apparently surplus to local needs, and the periodic over-use of limited social infrastructure. The theoretical equilibrium situation where labour is re-allocated between surplus and deficit areas rarely occurs in practice: local labour markets do not interlock and clear in the manner desired.

These local non-stable populations are in some important cases technologically determined: this is particularly evident where industrial economies of scale predominate, requiring huge inflows of migrants for production. The periodic labour surplus that results as the migrants' children in turn arrive on the local labour market is in a certain sense a 'relative surplus' in that it necessarily follows from the economies of scale of industrial capitalism. This terminology may sound familiar, but the phenomenon does not coincide precisely with any of the Marxist concepts of 'relative surplus' population (Marx, 1967). The type of 'relative surplus' that Marx referred to as the 'floating population' was determined by the needs of capitalism to cope with the phenomenon of the business cycle in such a manner as to avoid periodic rises in wage costs. In the case we describe here the 'surplus' arises because investment cycles and demographic generations do not coincide at the local level. That would matter little if investments were not so huge; but in certain industries they have to be huge to reap the economies of scale that will maximise private profits or state surpluses. In that sense the demographic outcome is the inevitable result of the requirements of capitalist industry given the technological state of the art. It impinges both on social infrastructure like housing and on future employment.

This then is the hypothesis we will explore, with the steel industry as the principal focus of our exploration. Our objective will be to examine how the industrial-demographic relationship might differ under conditions of changing technologies on the one hand, and varying demographic parameters on the other.

Demographic Consequences of Economies of Scale in Iron and Steel Production

The home of modern iron ånd steel making was the north-east of England in the nineteenth century. The main determinant of that particular location was the presence of the Cleveland ores. The major production centres that became established from the middle of the century were in the vicinity of Middlesbrough, but the site that parallels the Indian examples of green-field plants such as Jamshedpur, Durgapur and Bhilai, exemplifying in an extreme way the perpetuation of an industrial monoculture that lasted until 1980, lies further north at Consett.

Situated high on the Durham moorland on what was believed to be a rich source of ore, Consett was regarded as a huge plant in the early 1850s, yet its output was only of the order of 100,000 tons per annum. Tracing the size of the associated township over time is not easy, but the population of the Lanchester locality, described in the *Durham Chronicle* at the time as consisting of people dependant on the iron works numbered 20,000 (Wilson, 1977): hence the male workforce directly or indirectly employed in iron-making must have been less than 5,000.[1] The point to emphasise is this: the technology of industrialising England dictated local demographic features that were of a completely different magnitude from those in industrialising India of the 1950s; a local population of a mere 20,000 grew up over a couple of decades as a result of an investment in plant of the largest economic scale then known to the industry concerned.

The limited size and growth rates of these local populations in iron-making areas in comparison with the Indian experience of the 1960s reduced the planning problems related to green-field sites, and the population fluctuations that followed were of modest import. Furthermore, even quite moderate industrial growth rates were more likely to shift the population proportionally from rural to urban areas, given the slower rate of rural natural increase, than in India nowadays. Demographic pressure in the rural sector was much relieved, and migrants had to be sought from more distant places, a process no doubt encouraged by industrialists seeking cheap labour; high urban mortality would reduce the natural labour supply there, and even the local employment multiplier effect was probably smaller than in India today. Certainly the heavy industrial towns became centres of population diversity. Fifty per cent of Consett's initial labourers were Irish, and 'the influx of many new workers of diverse origin into the infant towns such as Middlesbrough…created social problems for which there was at first no adequate machinery' (Carr and Taplin, 1962, p.14).

From here we should turn to the origin of the steel industry in India. In 1911 the Tata Iron and Steel Company was founded in the district of Singbhum in Bihar, on a site in the jungle close to a vein of singularly pure ore. The township that emerged was christened Jamshedpur. It is significant that the technology sought was the best currently available in the world, obtained from the United States of America; the differing factor proportions of the two countries seem not to have been taken into consideration. This meant that technical expertise, physical capital embodying the advanced technology, and even the skilled and semi-skilled labour had all to be imported (Keenan, 1943). Ironically, however, the financial capital required was readily obtained from India. Despite the costs of importing skills and technology the plant was able to export steel competitively by 1912: the very high quality of the Indian ore must have been an overriding factor here (as indeed it has been more recently). By 1915 the plant produced 100,000 tons per annum, and by 1921 170,000 tons with a workforce of 26,000 (Sen, 1975). By that time Jamshedpur's population had reached 57,000, and it had become one of the largest towns in Bihar and Orissa. While being comparable in other ways to the most efficient steel plants anywhere in the western world it is clear that it was operating with a much larger workforce. By the 1950s new technologies and scale economies in the industry meant that the most profitable operation would take place in a million tonne (i.e. metric ton) plant: the necessary expansion was achieved at Jamshedpur by 1955 with a workforce totalling 29,000 (little more than in 1920); by then TISCO had become a fairly complex undertaking manufacturing a range of saleable steel items and we would estimate that under 20,000 were employed in basic iron and steel making.[2] By the mid 1960s capacity had been doubled to two million tonnes without a significant increase in the core-sector workforce (Datta, 1986): given the rapidly increasing labour supply as demographic growth accelerated following mortality decline, such a tight control on employment would have been unthinkable had the plant not remained in the private sector.[3] The city's population was heading for one quarter of a million by the 1950s and half a million by the 1970s.

In the late 1950s the Indian public sector undertook the construction of three new giant steel mills, with foreign aid and technical assistance, at Rourkela in Orissa, Bhilai in Madhya Pradesh, and Durgapur in West Bengal. The comparative picture is again instructive. For at Corby and at Consett in the United Kingdom renewal and expansion of iron and steel plant were also taking place at this time. Essentially the technology was the same as in India; indeed Britain designed Durgapur. But what makes a remarkable comparison is the size of the labour forces. At Corby in the 1960s British Steel (a public sector undertaking) employed approximately 6,000 men in their integrated iron and steel works (which probably represented 'overmanning'). According to the contemporary record Durgapur, Bhilai, and Rourkela were each designed to employ 7,500 (*Times of India*, 1959). In actual practice Hindustan Steel employed about 20,000 at Durgapur and 17,000 at Bhilai and Rourkela (Steel Authority of India Ltd., 1970); in each case the figures are likely to exclude contract labour (see note 3 below). By the 1970s Consett was using two blast furnaces and two L.D. converters to produce 1.4 million tonnes of steel billet employing only 3600 operatives; at Durgapur four blast furnaces and open hearth converters were used in a plant that was then rated at 1.6 million tonnes though output rarely exceeded one million; the workforce employed was by then about 30,000 (excluding the 7,000 employed at the alloy plant and excluding contract labour): what proportion of the total force was engaged up to the billet stage is not clear but must have been about 15,000 (including the service staff, on which note 2 below is illustrative). The difference in factor proportions is striking, and the demographic implications no less so.

The second point to note is that both Corby and Bhilai conspicuously failed to diversify. Employment in manufacturing amounted to 17,000 at Corby in 1966, with 6,000 or so in the integrated iron and steel plant and a further 6,000 associated with a tube mill (which mills steel from other plants in addition to Corby), thus totalling about 12,000 on the site. Here total employment in British Steel came to about 70 per cent of the manufacturing workforce. In Bhilai about 70 per cent of the manufacturing workforce were also employed by Hindustan Steel at that time. The whole of the manufacturing and services sector labour force supported a local population of about 40,000 in Corby by the mid-1970s and 39,000 at Consett; whereas at Bhilai the population had reached about 170,000 by 1971, i.e. four times the size of the United Kingdom populations at roughly comparable steel production levels.[4]

An identical technology has been adopted in these two economies, but owing to different circumstances of labour supply the demographic implications can be sharply contrasted. In the case where labour is relatively abundant the core industry itself carries more labour per unit of capital than in the case where labour is relatively scarce. Abundant labour has made it easier for industrialists to keep wages lower than they have become in the 'western' economies: at the same time scarce capital has tended (despite government subsidies) to keep the costs of equipment high. The tendency has been for Indian industrialists to use labour in place of equipment in order to keep costs down and preserve profits or surpluses. In the 'western' economies labour had become relatively scarce in the post-war period thus strengthening its bargaining power; where possible, industrialists used relatively cheap capital in place of labour. This is true in the supposedly profit maximising cases (contrast Jamshedpur with Consett in the 1970s, by which time there was probably little over-manning), and in the cases where employment was probably being generated for social or political reasons (contrast Durgapur with Corby in the 1960s). The industries that grow up to provide the local goods and services also carry far more labour in the labour-abundant economy. Labour in-migration rates are therefore far higher. In the labour abundant country family size is also larger, due to higher fertility without compensatingly higher mortality: this of course has ensured the greater abundance of labour. Hence total in-migration rates are higher too. These are reflected in a very sharp acceleration in the growth rates of the towns over a decade and a very large population at the end of the decade. When the technology was different so that optimal scale of plant was smaller, the acceleration in growth rates was naturally smaller. At a time when western economies were themselves more labour abundant than they are today optimal scale of plant was also considerably smaller. But the *net* effect of more abundant labour but smaller plant size was slower urban demographic growth and smaller cities at the end of these periods of acceleration than in developing countries today. Hence we believe that the experience in India in the 1960s is unique in history.

The argument we have outlined may be pursued—the epilogue is as follows. By the end of the 1970s it was not clear what the limit would be to increasing economies of scale in iron and steel manufacture. At that time the *minimum* efficient scale for an integrated plant was six million tonnes (Char, 1979). The next plant to be established in India was at Visakhapatnam, designed for a capacity of about three million tonnes in the first phase. The implications for local population growth would seem to be enormous. But economies of scale would clearly become labour-saving in character, and one cannot merely extrapolate from the cases of the 1950s and 1960s. Furthermore, technologies in iron and steel making are changing dramatically. We will return to this point at the end of the essay; the change is a crucial one, for it suggests that the demographic phenomenon we have just documented may not occur again.

Economies of Scale in Other Heavy Industries

Let us here consider to what extent we have taken a unique case for study in the iron and steel industry. Are there other industries with similar relevant characteristics of scale? For an answer we would expect first to look at the heavy engineering industries most closely related to iron and steel. These were an integral part of India's Second Five-Year Plan, and were also conceived on a 'grand scale.' A notable case was the heavy engineering complex at Ranchi. By 1971 there were two large plants in the district, with 18,000 employees recorded in total on their payrolls in the Census. The related population formed Ranchi Urban Agglomeration (including the new towns of Doranda and—a premonition?—Jagannathnagar), totalling 255,000 people, an increase of 82 per cent in a decade. At Durgapur the mining and allied machinery plant employed 8,000. Another important engineering plant in the area is the Chittaranjan railway workshop, which dates from the 1950s, and whose labour force had reached 5000 by 1961; this resulted in a local population growth of 79 per cent over a decade in a new-town built for the works, reaching 29,000 by 1961 (much of the multiplier effect being felt in nearby Asansol). An older railway manufacturing and repair factory is found at Jamalpur; employment here was 10,000 according to the 1971 Census.

   These are select examples from the eastern region of India. Taking the country as a whole we find that in 1971, among the factories that were recorded as employing at least 5,000 workers, the average employment per plant was 7,000 in machine-tool manufacturing, 8,000 in transport equipment, 12,000 in electrical machinery, and 11,000 in basic metals (which includes iron and steel plants; Table 4.1). At about the same time in the UK, West Germany, and the USA, mechanical engineering factories employed on average between 1,000 and 2,000 workers per plant, electrical engineering between 2,000 and 3,000, and from 4,000 to 12,000 in iron and steel production, these figures being from the upper quartile range of the employment size distributions (Prais, 1983) (Table 4.2). The rank order is roughly the same though the numbers are different. However, in Europe and the United States the motor-vehicle, ship-building and aerospace plants enjoyed much more significant economies of scale to judge from the same data source, giving larger workforces per plant than those in iron and steel. India had yet to penetrate these industries to any significant degree.

During the 1970s a change began to take place in the Indian industrial structure. Arguably this related to rigidities in the income distribution, such that the focus of consumer demand shifted upwards to the richest decile; this coincided with incipient change in the role of the state and public sectors, and perhaps in the ownership pattern of industrial capital itself. The issue is clearly important, controversial, and complex. What concerns us here is that a switch of emphasis occurred towards the industrial products based on the chemicals and petroleum industries, and away from coal and iron. Would this trend have implications for economies of scale and for local demography also?

In 1971 according to the census the largest chemicals or chemical products plant in the country employed nearly 13,000. The next largest was below the 5,000 mark. These figures are still not quite comparable to those of the steel industry but the margin of difference is not large. European and American experience was extremely divergent in these industries (indicative of the new and rapidly changing technologies). West Germany, which probably enjoyed a young capital age-structure and hence may have had industries that were reasonably representative of minimum efficient economies of scale, produced chemicals in plant that was larger by a small margin than its steel plant. In India the largest downstream plant in the chemicals industry (i.e. in rubber, plastics or petroleum products) employed only 6000 men in 1971. By the 1980s some truly giant investments were being made in this sector, as for example in the fertiliser plant at Thal Vaishet near Bombay, which was claimed at the time to be the largest in Asia. Employment is reported to be a mere 1,500; capital substitution for labour is clearly dominating the effects of economies of scale.[5]

In the current economic climate, and given changes in the economic strategy referred to above, we find an industry of particularly rapid growth is petroleum refining. Here again capital substitution is dominant. For all the difficulties involved in valuing capital across industries, we can fairly confidently assert that the petroleum refining industry is more capital-intensive in India than any other industry so far described. In Europe and the USA the upper quartile had an average of between 1,500 and 2,000 employees (depending on the country), with capacity between 10 and 15 million tonnes per annum. In the early 1980s the largest Indian plant at Trombay (close to Bombay) had a capacity of 5 million tonnes and employed about 1,000 men. Expansion in this field has very few demographic implications for the locality (and not least if the locality consists of a city the size of Bombay); at the same time it implies very little transfer of population from rural to urban areas.

To complicate the picture there are far less powerful economic forces attracting the heavy chemicals industry to green-field sites than in the case of iron and steel. Industrial or commercial economies of agglomeration (that is to say economies external to the plant) are likely to be of greater importance. These large industrial investments affect the local demography rather less if they are attracted to already developed industrial localities, like Bombay or Poona.

Implications of Economies of Scale: Social Infrastructure

To help focus the problem now under discussion we will outline the basic demographic scenario. The heavy industrial towns are distinguished by the very rapid growth in their labour forces. The first numerically large component in that growth is a category of wage labour which, after an initial period of living alone in the city, will quickly form a small household, usually bringing into the town not only a wife, but up to two children already born to the couple; within the first decade of the husband's residence in the town a third child will be born.[6] The point to note is the speed of growth, and the fact that the single-member households are only temporary.

In association with these migrants two other large population movements are found. The first (which actually predates the core labour migration flow just described) consists of construction labour, about which we have said very little, except to note in passing that, in the case of the steel towns at least, such labour is not eager to move out. Of greater numerical importance (eventually) is the second population movement associated with the core migration; this is the rapid formation of what I have termed a 'multiplier' population, namely a migration stimulated by the prospects of supplying the goods and services demanded by the core labourers and their families. Again there is rapid family formation, and a proportion (especially the poorest) migrate as a complete family. Again the striking characteristic is the speed of growth. In five years Rourkela grew from scattered villages to a population of 90,000; in the next decade it grew beyond the size of Middlesbrough (which took 120 years to get there); by 1981 the population was a third of a million. Durgapur had reached a quarter of a million similarly within 25 years.

Before the mills begin to deliver steel (or whatever) a huge investment is required in social infrastructure to meet the needs of this population. This is a net investment in the economy. The rural dwellings from which most of the migrants come are either occupied by other members of the family or allowed to decay. Ownership of the new urban dwelling would thus be a net capital expenditure for the migrant and direct purchase is out of the question. These very heavy capital expenditures required over a ten to fifteen year period in the life of these rapidly expanding towns have to be born by the state or central government. This is simply because of the coincidence of the requirements of a large number of newly migrant labourers to a green field site, where there is no pre-existing local population with savings to invest in municipal construction. The problem is that the whole population *and the industry* are at the same point in their collective life-cycles at that location—namely the beginning. Hence both will tend to be net borrowers.

The other problem stems from the rapidity of growth. Even if the need for heavy social infrastructure be met, considerable skills in organisation are also required to ensure optimal location of housing and other amenities over time. In the case of heavy industrial centres it would not be premature to plan for a city of a million when the current population is one hundred thousand. With hindsight it is difficult to believe there were those who argued that Rourkela would stabilize (whatever that might mean) at a population total of just that—100,000! The pure planning problem is of course eased with a better distribution in the pattern of demographic growth, in that the absolute volume of migration is the same but the phasing over time is different (Appendix A). The economic problem is also eased somewhat in that there will be more of a coincidence at the local level of net borrowers and net lenders.

We proceed here to a brief comment on planning practice and policy with particular reference to housing, in the newly expanding industrial towns. The first thing that strikes one is the lack of a co-ordinating body in the new-town areas. None of the steel towns had municipal councils, nor initially did any of them even have a development author ity. The planning was done by the public sector industry in conjunction with a 'Notified Area Authority'. The industry planned only for its core-sector industrial workforce (considerably underestimated in number as we have seen); the planners did envisage a 'civil township' of the order of one hundred thousand population, but the administrative system did no more than lay down guidelines in the form of zoning for future land use. A striking example is the case of Durgapur, where, despite the eventual setting up of the Asansol Durgapur Development Authority, the official construction of a central business district did not begin until ten years later (1971), by which time an unofficial business area had already been established on land zoned for residential purposes only (Sivaramakrishnan, 1985).

As regards housing, the public sector industrial companies were eager to retain their core workforces; this requirement ensured the provision of adequate housing for employee and family. In the steel towns the cost of housing worked out at about Rs. 15,000 per unit plus a further Rs.6,500 for land development and provision of public utilities; this was for a workforce whose average incomes were Rs. 3000 per annum. The subsidy involved worked out at approximately 80 per cent in practice. Surveys carried out in Bhopal and Bhilai industrial townships found that only about 40 per cent of the core industrial labour forces were, in the event, provided with public housing. This relates to the severe underestimate of the size of those workforces. In Durgapur, as no doubt elsewhere, the Notified Area Authority provided some facilities (but no housing) outside the main township area for some of the overflow.

Regarding the remainder of the city's population, virtually no provision was made at all.[7] Hence a very sharp duality emerged between the planned and unplanned sectors of the city, and their corresponding populations. The low level of facilities was experienced by a population far beyond those who could be described as the poorest, though they tended to be poorer (Crook and Malaker, 1988). But it was not of concern to the public sector industry to provide amenities for the 'multiplier' sector workforce, while the latter's own employers (if they were not self-employed) regarded this workforce as dispensable (a pool for hiring and firing), and hence were unconcerned about the risks of low productivity that might result from poor health or general domestic dissatisfaction. A study by the author indicates that poorer health in the family was indeed the result (Bapat, Crook and Malaker, 1989).

Implications of Economies of Scale: Employment of the Second Generation

The second major economic problem that occurs with the establishment of a large population on a green field industrial site lies dormant for some years, and hence will hardly cross the planner's mind. Nor is it of concern to the core sector industrial employer, for whom the time horizon is limited. I refer to the emergence of unemployment among the 'second generation' labour force (as we shall call the sons of the migrants). The precise manner in which this happens, and its relative severity, will depend on various demographic and economic parameters operating in different societies and localities (and the simulations in Appendix A may be found illustrative here). But the central point is that it must follow from the demographic conjuncture that is the direct outcome of the economic requirement to maximise the technologically determined economies of plant size. We will illustrate the scenario to make the point, basing our simulation on the experience of the steel plant at Bhilai in so far as this is revealed in the census data.

We can reconstruct what happened after the arrival of the first 15,000 or so steel workers (that is operatives and service workers for the Hindustan Steel plant, not including construction labour, nor indeed contract workers) as follows. First we note that Bhilainagar grew from 86,000 to 174,000 in the decade 1961-1971. However, the workforce in steel production increased to only 25,000 according to the census and the Bhilai Steel Plant records (BSP, 1984), and the total male labour force rose from 46,000 to only 57,000. What is more we estimate that only 8,000 to 10,000 male migrants joined the labour force during that decade (including male family members of the original workforce). In other words nearly 90 per cent of the 100 per cent decadal demographic growth came from 'natural' demographic forces: given the structure of the population that had amassed between 1956 and 1961 the rest followed naturally on the road to demographic stability.

But demographic stability takes nearly 100 years, not a mere decade, before it achieves its recognizable characteristics such as its unchanging age structure (Lotka, 1939; Coale, 1972). Hence our second rather surprising conclusion: to Bhilainagar was transposed a segment of the quasi-stable rural population with *very little interruption* resulting from the process of migration. Most of the men were already married, and their wives followed soon after their husbands' arrival; they brought their existing children with them and continued their child-bearing. Hence by as soon as 1971 a familiar age pyramid was restored in Bhilai, similar in structure to the rural population except for its youth-fulness (few migrants being aged over 45) (Figure 4.1). The sex ratio of 1961 that seemed so adverse to the prospects of fast natural increase turned out to be a false clue to the future. For in ten years the sex ratio had reached 1.25 (males to females). I estimate the birth rate to be around 42 per thousand in Bhilai in 1971 (which is remarkably close to what it would be in the stable population). The interruption caused by the process of migration was hardly sufficient to lower the fertility of the couples concerned. Although we do not go into the detail, our Fig-ure 4.1 illustrates a similar process at Durgapur.

If we concentrate now on the sons born between 1961 and 1976 to the male migrants who had arrived by 1961, the phenomenon described above gives rise to a quinquennial addition to the labour supply (i.e. those entering at ages 15-19) rising from 9,000 (in 1971-6) to 14,000 (in 1981-6). This is a dramatic increase, for the male population under 15 years of age in 1961 had only been 11,000 (implying about 2,000 or 3,000 labour force entrants over five years). The absolute numbers involved are very large too. But in addition, if our projections are substantially correct, around 11,000 new migrant males seem to have entered the township between 1961 and 1971 (over 5,000 in each quinquennium).

By 1971 the town of Durg and the new-town of Bhilainagar had coalesced, so that most data from the 1981 census refer to the whole urban agglomeration. This had grown by 84 per cent from 1961 to 1971, and it grew *again* by 100 per cent between 1971 and 1981. It seems that the local economy was buoyant in that it could draw in new migrants. But could it also employ the second generation from the previous cohort? There might be a contention that it could, supported by the evidence from the census on age-specific employment in manu-facturing and repair activity in the urban areas of the district (data for the urban agglomeration being unavailable): close to 60 per cent of adult males were so employed *in each age group* by 1981 (Table 4.3). There seems to be some similarity with the steel town of Durgapur here, but with a lower proportion holding in Burdwan district than in Durg. How-ever, Durgapur urban agglomeration had begun to grow more slowly in the 1970s (51 per cent over the decade); this is not surprising as increase in core steel-making capacity did not reach more than 1.6 million tonnes. But it is worth noting that whereas over 60 per cent of the workforce aged 35-50 were employed in factory manufacturing industry, the proportion diminished to between 20 and 30 per cent for the younger age groups entering the labour force in both cities (Table 4.4). The sustained proportions of 50-60 per cent in manufacturing and repairing in the younger age groups in the respective districts of these steel-towns may reflect a relative expansion of small workshops which is absorbing the younger generation (and the higher participation shown by the Census in household industry at the younger ages would support this); in such establishments neither pay nor conditions of work are so good. An alternative explanation of the sustained employment may be the effect of contracting out manufacture and especially repair work:[8] in the latter the census investigators may have apportioned a number of such workers to the services sector, rather than to manufacturing industry.

In Bhilai the problematic effect of massive growth in the labour force at age of entry has been postponed by the sustained rapid growth of the local economy especially in expansion of manufacturing in 'downstream' metal-goods (which did not happen at Durgapur). However, the core-industry remained iron and steel making. This employed 15,000 operatives in 1961 according to the census and the Bhilai Steel Plant records (BSP, 1984) (the true figure being probably about 20,000), and 25,000 by 1971 (the true figure being almost certainly over 30,000), while capacity had increased to 2 million tonnes. By 1977 two and a half million tonnes were being produced annually. The employment elasticity implicit in the output figures above is clearly well below unity. This implies diminishing employment potential in the face of increasing supply *in so far as* the sons of steel workers expect to follow in their fathers' footsteps.[9] We have seen how the age-specific employment rates imply adequate absorption within manufacturing: the same data, however, reveal diminishing absorption within basic metal processing with 45 per cent of 40-49 year-old industrial workers so employed compared with only nine per cent of 20-24 year-olds (Table 4.3). Of course some occupational mobility from generation to generation has to be expected, though it is well known that steel workers' sons do aspire towards the secure and well-paid employment in that industry. The more serious problem occurs when the downstream manufacturing investment ceases to expand at the same rate as the supply of the second generation in the locality. The 'demographic momentum' (i.e. the potential reproductive capacity of a young population) continues in full force at Bhilainagar, with about 37,000 women in their child-bearing years at the start of the 1970s (and potentially another 20,000 joining the men already there). In the urban agglomeration of Durg-Bhilainagar there were over 50,000 women of child-bearing age at this date. The greatest strain on the employment capacity of the town falls in the 1980s with between 15,000 and 20,000 new entrants to the labour force per quinquennium; the lower bound of the possibilities will occur (despite demographic momentum) if fertility per couple among core-sector workers has fallen (which is not implausible to judge from our survey in Durgapur (Crook and Malaker, 1988).[10] It is not until the end of the 1980s and start of the 1990s that major retirements of the first migrant labour cohorts occur easing the labour market squeeze.

A well-targeted expansion programme toward the original goal of 10 million tonnes could have provided all the employment needed at Bhilai though the time-phasing becomes crucial if the labour demand is to interlock precisely with the natural demographic supply. Table 4.5 attempts to reconstruct the interlocking of the investment cycle with the demographic cycle from 1960-65 to 1980-85. The first two rows refer to operatives at the worksite only. Between 1959 and 1965 about 18,000 of the migrants were absorbed as operatives. This coincides with a participation of about 39 per cent in the iron and steel industry. Between 1965 and 1970 young men emerged from those same households seeking employment: and as ingot steel output was increased to two million tonnes several thousand new jobs were created. Clearly they were filled by a combination of new migrants and the sons of those employed in the township and other services (or even as construction labourers). From 1970 to 1975 about 3,300 sons of the steel workers will have been ready to enter the labour market (Table 4.5). By now the demand for operatives is slowing down, and only 5,000 new workers were required to bring output close to the 2.5 million tonne level (achieved by the early 1970s). There were no further investments to come on stream in the period 1975 to 1980, though construction for the four million tonne stage was begun. Hence the considerable slowdown in recruitment, to 1,000 new operatives only (Table 4.5). The supply in the form of the sons of the current operatives rose substantially, however, to 5,200 according to our estimates. This indicates the failure of the investment and demographic cycles to interlock. It is possible, however, as we have noted above, that fertility may have declined as the core labour force discerned the advantages of small households: the acquisition of good qualifications to equip them for secure and relatively well-paid jobs as operatives with the Steel Authority no doubt seemed the wisest strategy for the sons of the steel workers, and this was a costly strategy for households: small families probably became the order of the day, as they did at Durgapur (Crook and Malaker, 1988). Besides, the cramped living quarters provided by the Steel Authority would not have encouraged family formation (to judge, at least, from the author's observations at Durgapur). If fertility had declined by 40 per cent by 1961-65, these new entrants will have numbered about 3,000 (not 5,000). There still remains a mismatch, however, between supply and demand in 1975-80 (Table 4.5).

By the early 1980s the foundations were laid for the expansion to four million tonnes, and a new plate mill went into trial production. Altogether for this period 4000 new recruits at the works are recorded by the Bhilai Steel Plant records. This coincides with a supply of second generation steel workers that has now reached its peak of 5,500, or just over 3,000 if we assume fertility to have fallen. A better match between the demographic and investment cycles is occurring again therefore as we move toward the four million tonne capacity stage (a match probably helped by some regulation on the demographic side). The investment planners will not have planned to ensure an equilibrium in the two cycles: indeed an excess supply of labour will have been welcomed as a relative surplus to maintain a downward pressure on wages, especially those of the non-permanent staff.[11]

Conclusions and the Future

We do not intend here to review the various strategies that might prevent or mitigate the problems of heavy industrial towns that we have described; these are discussed elsewhere in this book. The approach of this essay is positive rather than normative. One sentence will summarise its contention. The logic of industrial capitalism is such that in heavy industries like steel, mammoth plant created mammoth local labour forces in India with an inevitable demographic outcome: rapidly growing large cities with an initially unstable age structure and rapidly growing second generations. The social infrastructural requirements that this logic implied were *partly* met by industrial capital that regarded them as an inevitable cost of the programme to conserve core human capital. The second generation labour surplus also implied by the logic of the process, but following after a period of time, was regarded as neither advantageous nor disadvantageous to the programme and therefore did not enter the accounts so to speak. It could only be of benefit to capitalist development if investment was to follow in the same cycle as the demographic. That could only happen by chance as relatively short time horizons prevented it being part of the planning process. Finally one might add that the logic of this process has been mimicked in the socialist countries also (as for example at Maanshan in China): only the strategies to deal with the consequences may have been different.

Earlier in this essay it was remarked that, in the case of iron and steel production, there was no clear limit to the technical economies of scale, with an output of ten million tonnes being currently the best practice (as for example in Japan). However, in the 1980s a wholly new production technology began to be adopted in iron and steel making which may in the long run render the argument of this essay merely of historical interest, at least as far as steel is concerned. The increasing shortage of coking coal has encouraged the reduction of iron ore by means of natural gas. This process is subject to smaller economies of scale. Direct reduction kilns operate at lower temperatures than blast furnaces, producing sponge iron rather than liquid iron, and hence the furnaces require far less down-time for repair. It was the latter requirement that was the crucial reason for having massive blast furnaces, since quadrupling the volume only implied doubling the inside surface area which had to be periodically relined. A second reason for expecting change is that steel-making relies increasingly on scrap rather than iron ore, making proximity to existing cities more attractive than green-field sites; it follows from this that several small plants making steel in electric arc furnaces are more economical than one large one. So for the time being at any rate it may be that the phenomenon we have described will not recur in the Indian steel industry, which in the past was its most egregious host. This essay, however, has documented a principle at work. Empirically we will expect to see it elsewhere and again.

NOTES

[1] The population of Lanchester was recorded as 22,338 in the 1861 census, with males aged 20-60 numbering 5393. Consett works is said to have employed between 5,000 and 6,000 men. By 1864 the 18 blast furnaces at Consett had a capacity of 150,000 tons of pig iron per annum; by 1875 this was the largest iron plate works in the world (Richardson and Bass, 1965). The population of Lanchester grew accordingly by 103 per cent over the 1841-51 decade, and 41 per cent over the 1851-61 decade.

[2] It is fiendishly difficult to standardize the data on labour forces across different plants and in different countries. Production, maintenance and service workers are involved in basic iron and steel making (in blast furnaces and open hearths for example) in all integrated plant, the end product being blooms or billets. Most integrated works also include mills

for making plate, rail, wire or more complicated sections. It is usually difficult from published data to exclude those involved in the latter processes from one's totals, but as far as possible we have attempted to standardize by excluding the more specialized plant (such as the Durgapur alloy mill). The following breakdown of labour for Rourkela is informative (and also illustrates the problem). In 1962 there were 12,533 permanent staff and 3,565 temporaries (excluding contract labour). Of the permanent staff, 3275 were engaged at the coke ovens, blast furnace, and steel melting plants, 3237 were engaged in the rolling, strip and plate mills, and the remaining 6021 were in the service and maintenance sectors, such as engineering, traffic, and research.

[3] The example of the public sector plant at Rourkela was to make an interesting illustration on this point. At the two million tonne stage its total workforce became 40,000 with an additional 8,000 contract workers (Sengupta, 1985), compared with Jamshedpur's 30,000 plus up to 10,000 on contract (Datta, 1986).

[4] It is not very fruitful to try and estimate employment multipliers from these data as we do not know enough about import of consumption goods to the localities in question. But the effect of creating a core (steel) sector job seems to have been about one half of an additional job in the local urban economy in Britain and about three in India.

[5] Personal communication from D. de Monte.

[6] In Appendix A to this volume this demographic scenario is simulated.

[7] An exception was made in some cases (e.g. at Gopalmath in Durgapur) for the villagers dispossessed by the industrial development, for whom a resettlement location was found and some basic infrastructure laid out. It was also assumed that much of this population would be temporary, relating to the construction requirements; estimates of the size of these construction gangs run into very large numbers, averaging 40,000 for a steel plant. Not all proved to be temporary.

[8] Both at Durgapur and Bhilai the public sector steel plant took to contracting out for supplies on an increasing scale. For instance, in 1979, 34 industries around Bhilai were recognized as ancillary to the steel plant (by 1984 there were 40) (BSP, 1984). In this context it is worth noting that the total registered employment by the Steel Authority of India at Bhilai (including registered construction workers) rose from about 50,000 in the mid-1960s to about 65,000 by the mid-1980s, whereas labour force growth (including the new migrants) in Durg-Bhilainagar significantly outpaced this demand.

[9] In a seminar on Rourkela, trade union members advanced the following proposal: that 'vacancies arising out of superannuation etc. should be filled up by providing employment to the dependents of the superannuated workers as per the recommendations of NJCS agreement of 1983 so that settled families of retired employees are not put to any hardship'. The problem of the job-seeking second generation was firmly on the agenda, now that the problem, so clearly predictable, had actually arisen (ILO-ARTEP, 1989).

[10] See also Chapter VI, note 8, and Table 6.5 herein.

[11] According to press reports the training of senior executives at Durgapur and Rourkela steel plants now includes 'organizational structures' to deal with the need for 'a new cadre to fill the critical gaps left by a large number of retirements in 1993' (*Economic Times*, 24.8.90); this is an explicit indication of the recognition of the demographic dimension, late in the day.

Table 4.1
Scale of Manufacturing Plant in India, 1971

Industrial Sector	Average Number of Employees at Plant employing at least 5000	
Basic metals	11307	(a)
Chemicals and products	12889	
Machinery and machine tools	7225	(b)
Electrical machinery	12537	(b)
Cotton	6716	(b)
Jute	6034	(b)
Textile products	7050	
Transport equipment	8452	(a)

Notes: (a) calculated from upper quartile (b) calculated from upper decile.
Source: Government of India, Ministry of Labour, Labour Bureau, *Statistics oj Factories.*

Table 4.2
Scale of Manufacturing Plant in the United Kingdom and the United States in the 1970s

Industrial Sector	Average Number of Employees in Upper Quartile	
	United Kingdom	United States
Ferrous metals	7200	12000
Chemicals	1540	1270
Machine tools	550	470
Electrical engineering	2310	3200
Textiles	530	880
Man-made fibres	3730	6000
Motor vehicles	7200	12000

Source: S.K.Prais (1983).

Table 4.3

Age-specific Participation by Male Labour Force in Production
Occupations and in Metal Processing, Urban Areas of Durg District,
Madhya Pradesh, and Burdwan District, West Bengal, 1981

Age	Percentage in Production Activity		Percentage in Metal Processing	
	Durg	Burdwan	Durg	Burdwan
15-19	61	51	2	2
20-24	61	52	5	4
25-29	58	52	9	7
30-34	59	54	12	9
35-39	64	59	15	12
40-49	68	58	23	13
50-59	60	52	17	10

Source: Government of India, *Census of India 1981*, Economic Tables (Table B20).

Table 4.4

Age-specific Participation of Male Labour Force in Factory
Manufacturing at Bhilainagar and Durgapur, 1981

Age	Percentage Participating	
	Bhilainagar	Durgapur
15-19	25	22
20-24	33	32
25-29	37	41
30-34	38	56
35-39	46	65
40-49	63	65
50-59	51	49

Source: Government of India, *Census of India 1981*, Economic Tables (Table B3).

Table 4.5
Labour Supply and Demand Cycles at Bhilai:
Male Operatives at the Steel Plant

	Five-year Periods				
	Upto1965	65-70	70-75	75-80	80-85
Original migrants	18000	–	–	–	–
Sons of migrants	–	1800	3300	5200	5500
New jobs for operatives	18000	7000	5000	1000	4000

Source: Original migrants and sons of migrants are from the 1961 and 1971 Censuses and author's calculations (Government of India, *Census of India, 1961, 1971*, Madhya Pradesh, General Population Tables, State Primary Census Abstract); new jobs for operatives are from Steel Authority of India data (Steel Authority of India Ltd., 1984).

Figure 4.1
Age Structures in Rapidly Growing Steel Towns

BHILAI 1961
(population: 86,116)

BHILAI 1971
(population: 174,370;
decadal increase: 102%)

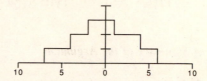

DURGAPUR 1961
(population: 41,696)

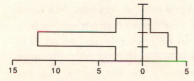

DURGAPUR 1971
(population 206,638;
decadal increase: 393%)

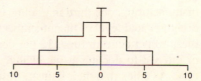

Note: Horizontal axes give percentage distribution standardized for five-year age groups;
four broad age groups are calibrated on the vertical axes: 0-14, 15-34, 35-59, and over 60.
Males are on the left, females on the right. These diagrams are for illustration only and are not
drawn strictly to scale.

Source: From data provided in Government of India, *Census of India, 1961, 1971.*
General Economic Tables (Table B2).

Chapter 5

INDUSTRIAL CONCENTRATION
AND INCOME SECURITY:
THE CASE OF THE SINGLE-INDUSTRY TOWN

A Summary of the Argument

This is a theory of two overlapping (or intersecting) sets. It is also a
theory of economic, social, and moral insecurity. Its unit of discussion
is initially the locality—city, town, or village (though it is an empirical matter to establish how appropriate a choice of unit this is). We
take the industrial city for illustration first.

The first set represents all the jobs in the city. In broad terms it
might be divided up with one third of working men and women in
manufacturing, one third in trading, and one third in services. At a more
disaggregated level of manufactures we might have five per cent in textiles, 10 per cent in food, 12 per cent in basic metals, and so on. It is
of interest to document the change in this distribution in a city over
time. As cities find their industry restructuring, the distribution changes
(so usually does the size of the city, but that is not immediately important). Some cities that are founded on a single industry, with most of
their manufacturing workers concentrated within it, at first experience a
highly skewed distribution of employments, which develops and diver-
sifies over time, tending towards a more normal or even uniform shape.
That is in every sense a happy development (provided it is also an
economically efficient development) and we shall not follow it up
further. Other cities do not develop in this way. Their employment
distribution remains skewed: one industry continues to dominate. We
shall call such cities 'industrial monocultures'. We chose the metaphor
carefully for it is meant to carry with it the connotations of a large agricultural area under a single crop—connotations of vulnerability.

Now consider vulnerability in a more extreme case. To fix ideas
more sharply, consider a coal-mining town. The town fails to diversify
because the financial advantages of the site only favour the mining of
coal, not, say, the manufacture of bicycles. But the capitalist, public or
private, who owns the mine, need not be worried. His whole enterprise

is not vulnerable so long as he has other mines at other sites, and preferably other enterprises (in bicycle manufacture, say) in other places also. What are the sources of vulnerability? The world price of coal may change, rendering this mine, in financial terms, unprofitable. New technology may be developed elsewhere, again, for the same reason, to the disadvantage of the mine, and to the advantage of newer ones. The mine may become physically exhausted. In all these and other scenarios the majority of the mining workforce in this town lose their jobs, and, given the case of monoculture, so therefore does the majority of the town's workforce.

It is immediately important to note that the story does not, of course, stop here. The trades and services sector workforces in this mining town are employed very largely to meet the local demand of the miners (and of each other). When the mine closes there is an *ejection* of autonomous demand (to adapt the economist's terminology) and a 'reverse multiplier' sets in, a downward spiralling of demand for goods and services. In a more diversified town, with, say, half of the non-tertiary sector workforce in coal-mining and the other half in cotton milling the worst effects of the reverse multiplier could be mitigated. Theoretically all the trades sector people, like the shopkeepers, could take a cut in turnover when the mines closed, and wait until business picked up when new industry came to the town (textile machine manufacture perhaps). But in a single industry town this cannot be done: the shopkeepers have to close down too as their turnover is cut to zero. In the extreme the whole town heads for destitution.

This in itself we would regard as a social disaster. In a true sense of welfare economics it is also an economic disaster, to which a mass population exodus is no solution (except for the few people who may enjoy shifting their location). Here we want to focus on a different but related aspect of the problem. We return to the overlapping sets.

The second set relates to sources of transfer payments in the town or city. Apart from incomes from employment in the city's industries, the workforce and their families receive sources of sustenance from friends, from the community, from the city exchequer, from the central government, and so on; these may be gifts, or relief schemes, or loans, or whatever. All those transfers come from earnings or funds that have ultimately to be got from productive employment somewhere in the economy. The employees who are the ultimate *source* of these funds are themselves distributed in various industrial employments (just like the recipients). Their collection of jobs forms the second set.

Now we seem to have a way out of the monoculture and its vulnerability. The transfer incomes (as we shall call them) sustain the miners *and* the shopkeepers during the hiatus in their employment and revenue earning life. In western societies the transfers are largely derived from central government; in India they are largely derived from members of the extended family. But it is only under specific and favourable conditions that vulnerability is thwarted.

In so far as the transfer incomes have the same source as the direct incomes, i.e. come from the same economic activity or industry in the same town or city, the two sets will overlap. For example, if miners usually help out other miners when they are sick or out of work, the overlapping part of the sets represents mining employment in each. Since, in this example, the support of miner to miner only goes as far as the boundaries of the mining town, the overlapping space represents the coincidence of two identical sources. Direct *and* transfer incomes derive from the same economic source. Close the mine, and both direct and indirect sources of sustenance are severed.

Now let us take the story a step further. Suppose within the same monocultural town shopkeeper helps out miner and shopkeeper when times are hard, and miner helps out miner and shopkeeper, (and there is no income transfered from outside). The two sets completely overlap; the one is wholly contained in the other. In such a society the closure of the mine entails the starvation of the town.[1]

Theoretical Discussion at the Household and Industry Level of Analysis

Let us return to discuss this problem at the household level in some detail. A family may be composed of a number of earners of various ages. In India its major source of income is likely to be from one or two adults. In the latter case if both are employed as wage labour in the same industry, or, in a more extreme case in the same factory, the whole family is very vulnerable to the fortunes of that industry or factory. Security is not much enhanced in an alternative model where the main breadwinners are all employed in an industry providing goods and services for the core industrial labour force. A slightly better position is offered by the familial mode of industrial production, as work sharing is a possible solution provided that the average productivity in the family does not fall below average consumption requirements; but once again the closer the work is tied to a single industry (for example through contracting out to the household) the more risky the household mode becomes. Clearly what emerges from this is that the safest strategy is to diversify the family portfolio, as it were. To have two or more adult workers in two or more quite distinct industries would be ideal. To have one member in an industry as a wage labourer, and one self-employed or working in household industry in a less closely related sector, would be a moderately secure strategy (perhaps at the cost of a higher income than would have been enjoyed with both members as wage labourers in the same industry). Of course a single but high enough income will generate savings enough to carry the family through a period of unemployment. However desirable this may be it is a distinctly long-term solution.

In practice the choice is often limited, even in the Indian urban economy. What evidence there is seems to suggest that one secure and good income source is the first preference of a household, with, for example, the head of the household employed in a modern industry or in municipal service. If that is not possible, or if that source fails as a result of compulsory redundancy, then other family members also have to participate in the workforce in whatever employment they can find. Often women are pushed into employment by force of circumstance rather than choice in this way (Bapat and Crook, 1988; Rodgers, 1989). Though evidence suggests that income may be sacrificed for security, there seems to be no evidence of strategies being deliberately adopted to diversify sources of household income in the manner indicated above. For quite apart from the constraints of an imperfect labour market which may restrict certain castes from certain occupations, there is the additional constraint of a lack of industrial diversification, and hence choice of employment, in precisely those localities where such household strategies could be most recommended. You cannot easily diversify your portfolio in an undiversified world.

It is probable, however, that families do adopt some longer-term strategies for survival, conscious of the vulnerability of their current income source. Education of children is a strategy of this kind: this helps to promote social and spatial mobility so that the source of future transfer incomes into the household may become quite distinct from the source of earned income. But education has a current cost, both in material terms and in keeping children out of immediate employment; the poorest households cannot afford such an investment.

Whether or not households, let alone broader social groups such as caste, regional or ethnic minorities *consciously* adopt such strategies, it is worthwhile investigating whether in practice the household vulnerability problem arises. Are urban family income sources particularly concentrated, or more concentrated than they need be, given the range of opportunities? For if they are, then the failure of local industry to become diversified becomes even more socially undesirable.[2]

The Evolution of Indian Industrial Monocultures

Before turning to the micro-level evidence let us look at the macro-level evidence. A glance at the composition of manufacturing as initially 'monocultural' cities evolve will help establish whether the opportunities for household income diversification have remained limited or not. Let us take a glance first at the composition of manufacturing in the three major steel town districts of eastern India that were rapidly developed in the late 1950's Second Five Year Plan period, namely Burdwan in West Bengal (containing Durgapur and also the older Burnpur), Sundargarh in Orissa (containing Rourkela), and Durg in Madhya Pradesh (containing Bhilainagar). The heavy concentration of large units in the basic metals, machinery, and transport equipment industries in Sundargarh and Durg in 1961 is immediately apparent (Table 5.1). Burdwan is more diversified, including a large scale food and beverage manufacturing sector; indeed the district has a longer history of industrialisation. In Sundargarh other developments were under way by 1971, noticeably a large-scale chemicals sector. That could be a crucial diversification. Burdwan achieved a similar mix of heavy industry, including electricals. The expansion in basic metals was only slight in Burdwan, and we have remarked in an earlier essay how the failure of the steel industry to grow at Durgapur severely curtailed local employment opportunities for the sons of the migrants (Chapter 4). Very little new large-scale investment came up in Durgapur during the 1970s. In Sundargarh and Durg there was substantial subsequent expansion in the basic metals sector. The lack of diversification here contrasts sharply with that in a region such as Poona district in Maharashtra, which we have taken for a comparison (Table 5.1; Figure 5.1). Even by 1980 the largest single industrial sector in the urban district of Durg, namely basic metals, still accounted for 70 per cent of the manufacturing employment, 58 per cent in Sundargarh (site of the Rourkela steel plant), 45 per cent in Burdwan, and 46 per cent in Singbhum (site of Jamshedpur) (Table 5.2). The latter was by then 70 years old.

In analysis of this kind the question inevitably arises where to draw the boundaries of the local economy, given the ability of labour to migrate. How far beyond the town and city boundaries should our employment sets extend? Migration between Bhilainagar and Durg is a matter of daily commuting. Migration between Durgapur and Burdwan City or Asansol-Burnpur may be a more serious matter. Migration between any of these and Bhopal, say, is more likely to amount to a social upheaval: family income and social ties become weakened. It is true that inter-state migration took place when the steel towns were formed. But the rapid family formation indicated that the migrants expected to stay in their new location; and as they age their propensity to move again is diminished, so that the social costs of forcing them to do so increase. The point to be made here is simply that such a skewed industrial structure as some of these localities have continued to experience is known to be particularly vulnerable to local or macro-level change in comparative advantage, or in economic strategy, or even in state-level or national prosperity. The western world's experience is not irrelevant here. The British steel towns of Corby and Consett (discussed in Chapter 4) suffered just such an affliction, as did the coal towns of south Wales. When the sources of relief available to poor people are largely determined locally, as in India, the problems are at their most acute.

Strategies of Family Formation and Occupational Diversification: the Empirical Evidence

It has already been established in a previous essay that towns of different industrial character differ often quite markedly in their demography. Among the economic implications that this might have at the local level, we are particularly concerned to stress the effect on family welfare, family income, and the security of the family over time. To do this we need to investigate the household structures that lie beneath the aggregate urban demographic structures. Fortunately in the Indian case there are two sources of information from which ideas may be culled on the distribution of household types within the urban areas under study. The first is the simple distribution of household size from the census tables of 1961, and the second is the Labour Bureau study of industrial worker households in selected towns undertaken in 1958 (GOI, Ministry of Labour, 1968).

As a first and unremarkable generalization we can say that urban households are smaller than rural households in industrializing regions. This follows if urban households are fragments of rural families. The Labour Bureau data give evidence of such split families, and the census shows that the majority of lone men in the early days of the steel towns were in fact already married. The greater preponderance of households consisting of only one, two, or three members in urban areas is largely due to migration; the role of lower fertility is initially less important, though with time it may come to predominate. The rapid growth towns are also distinguished by the absence of older relatives: again this keeps the average size down. The towns of more moderate growth (or rapid growth towns that have entered a period of more moderate growth) have a more similar household composition to the rural households; and the slowly growing urban areas are almost the same as rural areas in this respect (their aggregate growth coming from natural increase alone).

The heavy engineering localities are immediately distinguished by one feature: the high proportion of single member households. These are predominantly male, and, as we have indicated in a detailed discussion of Bhilainagar steel town in Chapter 4 of this volume, their composition rapidly changes: between 1961 and 1971 the proportion of single-member households declines in such towns from around 20 per cent to around 12 per cent. In fact the migration data we examined (Chapter 3) suggests that individual families tend not to remain split for more than two years on average. Over the ten-year period these households accumulate children also, which helps to balance the sex ratio further. However, a degree of masculinity in the sex ratio remains; in fact all towns have a higher proportion of unrelated members of the same household and married brothers in their household compositions than in the rural areas.

It would be advantageous to be able to analyse the core industrial sector migrants separately in the hope of discovering something regarding their vulnerability, as families, to economic shocks. This we cannot do for the most rapidly growing towns since the Labour Bureau surveys do not include any; however they do include some of the heavy engineering and steel towns of an older generation which underwent some expansion in the 1950s, namely Asansol-Burnpur and Jamshedpur: the former grew by 78 per cent between 1951 and 1961, the latter by 50 per cent. Asansol's industrial workers come from remarkably small households: 33 per cent have single members, and 43 per cent between two and five members. One third of those single-member households consist of unmarried persons. The industrial worker households are more likely to be single-member than are the urban households in general. The explanation may be that industrial workers are more likely to be recent migrants, and the divided family is a temporary phenomenon. Jamshedpur on the other hand is sharply contrasted. Only 14 per cent of the industrial worker households are single-member; this could easily be explained by the smaller in-migration associated with the expansion programme. The 1971 census data seem to confirm these interpretations. Asansol had roughly the same proportion of single-

member households as did Jamshedpur by that date. Tentatively we conclude that the migrant in heavy industry seeks where possible to transfer
his nuclear family from its rural base. Perhaps this weakens the support
that could be obtained from this source in time of need. However, as we
shall see, the matter is more complicated than this.

We should now ask how far the migrant broadens the base of his
family's income from sources in the urban economy. Our data of course
restrict us to an analysis of income from occupations reported to the
survey and census investigators (Table 5.3). In the case of both
Jamshedpur and Asansol multiple sources of such income are rare in
industrial households. In 80 per cent to 90 per cent of such households
there was only one earner; this is striking in view of the fact that there
were only 14 per cent and 35 per cent of single-member households in
Jamshedpur and Asansol respectively as we have seen, and in a quarter
of the households there was an adult member present *in addition to* the
married couple. This is in line with survey data from elsewhere indicating that industrial workers are not inclined for their wives or school age
children to enter the labour market so long as they themselves are earning (Bapat and Crook, 1988). In Jamshedpur and Asansol industrial
workers earned at least the average income for the town, and over 60 per
cent of them were in the iron and steel industry, most of the rest being
in some branch of heavy engineering.

To put these findings in perspective we will proceed to look at
some contrasting types of towns. Keeping first to the industrial monocultures, we take a group that is subject to large fluctuations in growth
and prosperity, the coal-mining towns. The Labour Bureau surveys include studies of Raniganj in West Bengal and Jharia in Bihar. Both
experienced slow growth over the 1951-61 decade, the former with net
out-migration, and the latter with hardly any net inmigration. Both surveys consist almost entirely of miners. Once again single-member male
households are common (between 40 per cent and 50 per cent of cases).
Given the towns' modest growth rates these would appear to be permanently split households; indeed less than three per cent of the men are
unmarried. The other contrast with the manufacturing industrial towns
is that 15 to 20 per cent of the mining households contain additional
family earners also working in the coal town (in between five and eight
per cent of cases the additional earner is the wife). A combination of
lower wage rates (see Table 5.3), and probably larger families to support overall, including the rural part of the family, has compelled this
higher participation. In the two mining towns the additional household
earners are themselves nearly all employed in the mining industry also,

the women as surface workers presumably. This extreme vulnerability due to concentration of income source is illustrated by the tragic sequence of events that is sometimes reported when a series of mines are closed down: in the late 1970s there were accounts of starvation in some ex-mining communities of Bihar. Such an event would be less likely if miners had been in receipt of transfer incomes say from family agricultural holdings. But miners are typically from poor, probably landless, families: they are disproportionally from the scheduled castes and tribes. Furthermore in an average year around 40 per cent of miners' households report sending remittances back to their families (a higher proportion than in the steel and engineering towns): there is clearly little scope for an income transfer in the reverse direction, from country to town, if the urban part of the family is in distress.

Monoculture is not restricted to mining or heavy industry. By way of a contrast and comparison with what has gone before we take a look at two cases of textile manufacture in western India. The similarity between the cities of Ahmedabad in Gujarat and Sholapur in Maharashtra lies in their both being dominated by the cotton mills. The difference lies in the more modernized industry in Ahmedabad and the more backward industry in Sholapur; this is reflected in the corresponding growth and population structure of the two localities, with decadal increase in Ahmedabad being 37 per cent from 1951-61 compared with 22 per cent (implying no net migration) in Sholapur. Ahmedabad bears some resemblance to Asansol or Jamshedpur in having 21 per cent of its households consisting of single-members only, 77 per cent of its industrial labour households deriving their incomes from one principal breadwinner, and 91 per cent of its industrial labourers being employed in the textile mills (Table 5.3). Sholapur is a case of industrial stagnation in the 1950s. It has only four per cent of its households consisting of a single member (one third of whom are lone, probably widowed or deserted, women), and 45 per cent of them are over six members strong (compared to 22 per cent in Ahmedabad). Nothing changes by 1971; and these characteristics are similar to those in the district as a whole. This is indeed a case of the 'low-level' equilibrium steady state. The proportion of multiple-earner households is high at 47 per cent; and the principal earner can only expect to earn about one third of the occupational income of a fellow weaver or spinner in Ahmedabad (or of an iron and steel worker in Asansol or Jamshedpur). In view of the low wage it is surprising there are not more than 20 per cent of households with a second earner: the product demand factor is undoubtedly a constraint. Furthermore, the 1961 census data indicate that, of all the *household* industrial units recorded, 73 per cent are in cotton textiles (in contrast with only 50 per cent in Ahmedabad), an extreme degree of concentration of earning power in a single industry. The family and population structures in Sholapur suggest there are few remaining links with a rural economic unit, which, taken in conjunction with the evidence that less than two per cent of the industrial worker households claim to remit income out of the town, gives a picture of an isolated economy, with no scope for transfer incomes to alleviate economic distress. We seem here to be able to witness the full implication of a single-industry town.

Industrial and Occupational Concentration among Migrants: the Empirical Evidence

Our interest in employment diversification among migrants is twofold. First, as we have said, the major concern of this chapter is with the stability of income, which we have suggested depends on the mechanisms for transferring incomes from those with employment to those without. Generally speaking this will occur *within* families. But additionally it will usually occur *between* families that feel in some way a moral, emotional, or expedient responsibility for others. One of the strongest links here will be ethnicity; and a strong, if less strong, link will be place of birth. Hence migrant families of common origin, born in the same state for instance, may be expected to be supportive of one another in distress.[3] If migrants draw their incomes from diverse sources, the reliability of this form of security is enhanced; if they do not, then distress will be more likely to be concentrated, and inter-family income stabilization less easy.

Secondly we wish to assess what the strategy of migrants in this situation actually is. Do they themselves see the importance of diversification or not? It is after all possible that a conflict arises in that the best income or best security for the *individual* is to be found in particular industries or occupations. Hence migrants would aspire, through whatever links they manage to forge via their own kith and kin, to enter those industries: furthermore, the more of their members they can establish therein, the easier entry becomes. These atomistic advantages may appear to outweigh any collective advantage of diversified employment, whether perceived or not.

For some towns we can look at the evidence relating to migrants from particular origins (using the special census tabulations). In Jamshedpur it can be seen that the iron and steel industry employed 29 per cent of the local Biharis, but 38 per cent of the West Bengalis (a linguistically distinct group), and 39 per cent of the Orissans. In the Dhanbad Town Group, a coal-mining region, 37 per cent of the local Biharis were in mining, and as many as 49 per cent of Uttar Pradeshis. These are fairly clear indications of high degrees of concentration among particular migrant groups in town complexes that are already characterised by employment concentration in specific industries, and hence especially vulnerable to economic fluctuations. Specifying the locality of origin in further detail we find that in Dhanbad 43 per cent of the migrant workers from the neighbouring district of Hazaribagh are miners.

Comparing these observations with those from a textile city, Ahmedabad reveals a parallel case. Only 42 per cent of local Gujarati migrants are employed in the textile industry, but 66 per cent of the Uttar Pradeshis are so employed (a distinct linguistic group). The skewed sex ratio among the latter indicates that part of the family remains in the rural locality of origin (unfortunately we do not have data on district of origin from within this huge and economically diverse state). They may have land, but feel the need to supplement their income. The rural home will offer security in time of industrial distress, but the capacity of the village as a whole to perform that function is limited if there is a concentration of textile workers from a particular village. Case studies from the districts like Ratnagiri or Satara, which have traditionally supplied migrants to the textile industry in Bombay, suggests that this may be the case (Dandekar, 1986).

It remains to be seen how these migrant groups behave over time. We have only the evidence of those who have remained in the town in question—in some sense they may be referred to as the satisfied migrants. Census data limitations restrict us to looking backwards from 1961. Have these migrants become more concentrated in their employment characteristics over time or not? On the whole it would appear that they have. Production process workers in Jamshedpur (who will be mainly iron and steel workers) are nearly twice as likely to be found among those migrants who have been in the city from one to five years, as from among the more recent arrivals. This intensification is strongest among the Bengalis, who are already the most concentrated group. In the case of Dhanbad there is by contrast very little evidence of such increase in concentration. In the case of the miners specifically the reverse if anything is true. Not surprisingly perhaps, there is an attempt to get out of mining with time: it is after all an occupation which is both badly paid *and* insecure so that no question of a trade-off between the two arises. In the case of Ahmedabad we can distinguish between industry and occupation; but either way the overall pattern is similar. All the migrant groups tend to move towards jobs as production workers over time, whether in textiles or not. This movement applies no less to the Uttar Pradeshis, whose employment is more heavily concentrated in production work in the first place. What particularly distinguishes them from other groups is their initial degree of concentration. Indeed these migrants from U.P., of whom we noted the majority are in textile occupations (spinning or weaving), intensified their concentration in the textile industry over time: of migrant employed in manufacturing 65 per cent of those who arrived within the last year are in the textile industry, rising to 90 per cent for those of less recent arrival (6-10 years ago).[4] Of course these are historical data; the opportunity for similar behaviour in recent decades may well have diminished. The implication of such extreme concentration, not only in a single industry, but in a single occupation, has manifold ramifications. It seems to have particularly characterized the cotton textile industry within India, which is an industry whose future is none too bright. In the case of the Uttar Pradeshis, by carrying on family employment at two centres, at home and in a distant city, they may have perhaps maintained the diversification at the family level that we argue is so important, splitting their incomes perhaps between agriculture and industry. This, however, is at the cost of a settled family life. A study in Bombay reported how the migrants from districts of the Deccan who had a greater likelihood of having land were expected to survive the 1982 textile strike far better than the more frequently landless migrants from the Konkan (Dandekar, 1986). A contrast could be made here with the family employment in textiles in Sholapur, where the rural links seem to have been severed for good.

The Policy Issues

Our interest in the relationship between demography and industrial structure, which is an interlinking theme in these essays, has been prompted by a concern for some of the welfare implications of industrial strategies. A particular problem has been the insecurity of incomes obtained through employment in single-industry towns. It is well known that changes in industrial structure or technology may render a high proportion of the workforce unemployed, and a higher proportion in such towns than in more diversified locations. Is that observation alone an argument for encouraging (i.e. subsidizing) local diversification?

If there is a freely operating market, industries of a similar nature, or highly interdependent industries (in an input-output sense) will tend to locate in close proximity to one another to reap economies of production related to the physical fact of their mutual proximity (Mills, 1972). To attract wholly unrelated industries to the same area may require heavy subsidies continuing over time. To be sure it will often be easy to attract industries producing consumer goods at a rather low level of subsidy if the local population constitutes a market which is close to the threshold size at which the local industry would in any case become profitable. But then, as will be apparent, the social benefit of a subsidy for such a consumption goods industry will in this case be considerably smaller than that for any industry that exports most of its products beyond the locality.

How can any such subsidies be justified? Clearly the answer is not on any grounds of industrial efficiency. But if the state is involved in expenditures relating to unemployment relief then it is possible that the reduction in the need for such relief may more than compensate for the subsidy needed to promote diversification. We should perhaps here remind the reader of our argument in the opening section: in case of industrial closure, unemployment rates are higher in industrially concentrated localities not only because of the direct unemployment that results from closure, but also because the local multiplier puts more people *completely* out of work than in a more diversified local economy. The social return to a policy that attracts a second manufacturing industry, unrelated to the first and exporting its products beyond the locality in question, would be higher than the return from attracting an industry whose major function was to provide commodities or services for the locality itself: for the latter industry would itself add to the unemployment toll if the first industry went into a severe enough recession.

There is a further consideration here, arguably in the realm of the unquantifiable, but certainly of significance. A given level of unemployment concentrated in a locality may be socially more costly than the same amount of unemployment dispersed over a region. Localized depression may encourage crime and other social malaise, may discourage labour itself from retraining or seeking new job opportunities, or may discourage new investment from entering the area if entrepreneurs are sufficiently deterred by the environment of a depressed locality (and not sufficiently attracted by the low-wage potential precipitated by high unemployment). And we have already commented adversely on the social effects of forced out-migration of an older generation.

Many states in the developing economies of the world do not have direct unemployment relief. But this does not imply that the state is un-involved in interventions aimed at promoting local employment, such as employment guarantee schemes to counter the rural disaster of a crop failure. Urban disaster is no less real when the company town fails: the recent case of Dalmia Nagar in Bihar makes the point.[5] In India the national and state governments have been active in legislation and incentives regarding industrial location (Mackie, 1982); but they have failed to address the problem we outline here. Programmes of urban diversification should be already high on the agenda of a state that has become involved in programmes aimed at improving the income security of rural households.

Conclusion

We seem to have created an agenda for research rather than to have com-pleted any such programme. For we would like to know more about the incidence of overlapping sets of employments giving direct and transfer sources of incomes. The questions raised and discussed have a particular topicality in a world where so much industrial restructuring is taking place, resulting in the closure of old as well as the opening of new in-dustries. In many parts of the world, and not least the western world, the potentially undesirable welfare economic effects are being further exacerbated by the diminished role and finances of the state, reducing the scope for transfers from the gainers to the losers, from the broad to the narrow economic base.

As in our previous essay, we can see the problem here as a mis-match between industrial and demographic dynamics. The industrialist can diversify his portfolio – wind up one business and expand another with little loss: indeed the broad-based major Indian industrial houses are past masters at the art. The household cannot easily diversify its labour income in the same way: we have reviewed in this essay some of the possibilities and some limitations on this strategy in India. Perhaps more importantly we have to see the problem in terms of the differenti-ation of the urban household. Those with financial assets can diversify their capital, those with only labour resources have no capital to diver-sify. Those with surplus can live through a crisis, those without can-not.

There is perhaps a lesson for the victims of industrial restructuring, be they the Bombay or Sholapur textile workers, the Welsh or Bihari coal miners, or, one day to come, the steel men of Durgapur and Bhilai, like their bygone colleagues in Corby and Consett. For at such a time of potential siege it is crucial to keep the lines of communication open; to ensure that sources of material supply are various and their base diversified. This essay on overlapping employment sets implies a strategy for all social groups whose employment is industrially concentrated, and for the local communities that depend upon them. Minimize the overlap between direct and transfer incomes: seek your allies for economic support from far and wide.[6]

NOTES

[1] In the extreme, however diversified is a locality's economy, it may be possible to identify an afflicted group (perhaps typically an in-migrant group) whose range of direct and indirect incomes derives from a single source – say a declining textile industry in a particular quarter of a metropolitan city, or a plant hit by industrial disaster that blights a complete neighbourhood. Conversely, the geographical delineation of the locality may be less important in assessing vulnerability from concentrated income sources in certain circumstances: if for example members of a particular ethnic group up and down the country are concentrated in a particular industry, and help one another out in time of need, they are driven headlong downhill as a community if that industry gets competed out of business by overseas competition or whatever. The only advantage of the non-geographical concentration in this case is that the local multiplier effect is reduced.

[2] It would also be possible to investigate to what extent the demographic structure of the urban family, resulting from high or low fertility in conjunction with various patterns of migration exacerbates, or alleviates, this vulnerability. And if the strategies adopted are indeed conscious we could endeavour to see how the vulnerability problem is a constraint on demographic behaviour: for example high fertility may be a response to the need to diversify labour income resources.

[3] It is well documented that people of common origin tend to cluster spatially in the cities. Social networks of this kind are also documented in Gore, 1970.

[4] Unfortunately the census data do not allow us to follow through these migrant groups from census to census: the detailed tabulations made in 1961 were unique. Relying on retrospective data, in the way we are forced to, has its pitfalls; we cannot, for example, adequately control for the changing labour demand in particular industries as we go back over time.

[5] Dalmia Nagar, in Rohtas district of Bihar, provides an interesting but qualified illustration of our argument. This was not a single industry town in

the sense of being a steel or textile town, but its industries were under a single private sector ownership (Rohtas Industries Limited). When the company had to close down its operations at Dalmia Nagar in 1984, apparently through being unable to pay the electricity dues, nearly 13,000 workers were laid off. By 1987 the total unemployed in the locality had risen to one third of a million (ARTEP, 1989, p.86).

[6] At the time of going to press the provisional 1991 census papers were beginning to appear. The quite dramatic slow-down in the growth of steel towns is already apparent, a continuing indication of their failure to diversify substantially. Bhilai has the highest decadal increase at a modest 40 per cent, followed by Durgapur at 33 per cent (which is below the national average), and Rourkela at 24 per cent (a growth rate no faster than the rate of natural increase alone).

Table 5.1
Distribution of Industrial Establishments in Steel Town Localities and
in Poona, in 1961 and 1971

Industry	Sundargarh		Durg		Burdwan		Poona	
	1961	1971	1961	1971	1961	1971	1961	1971
Food	28	224	472	535	785	1595	1197	4180
					(17)	(19)	(4)	(3)
Beverage and tobacco	5	14	79	265	210	158	908	1003
				(4)	(3)	(1)	(7)	(4)
Textiles	54	238	550	728	386	1041	1841	2205
			(1)	(2)	(1)		(10)	(11)
Wood	49	97	173	207	167	–	404	1129
					(1)		(4)	(5)
Paper and printing	3	17	18	38	75	159	255	480
							(22)	(21)
Leather	22	12	64	97	45	72	356	264
								(1)
Rubber, petroleum, coal products	–	20	4	25	17	37	32	92
		(4)		(2)			(1)	(4)
Chemicals	8	46	21	21	37	125	141	174
		(10)	(1)	(1)		(2)	(12)	(9)
Mineral products	23	31	109	110	14	147	163	207
	(5)		(1)	(1)		(4)	(5)	(23)
Earthenware	2	–	–	–	45	–	–	–
Basic metals and products	17	71	89	89	202	365	736	892
	(2)	(17)	(11)	(10)	(11)	(15)	(11)	(20)
Machinery	–	29	33	38	63	121	200	539
			(8)	(8)	(1)	(5)	(9)	(48)
Transport equipment, repairs and miscellaneous	31	412	471	816	719	1710	1941	1723
	(5)		(12)	(2)	(6)	(9)	(4)	(25)

Notes: The data are for Districts (urban); figures in brackets are the number of
establishments with more than 50 employees.
Source: Government of India, *Census of India, 1961, 1971*, Orissa, Madhya
Pradesh, West Bengal, Maharashtra, Establishment Tables
(Table E3 (1961), Table E2B (1971)).

Table 5.2
Distribution of Factory Industrial Workers in Steel Town Areas, 1981

Industry	Sundargarh (Rourkela)	Durg (Bhilai)	Burdwan (Durgapur)	Singbhum (Jamshedpur)
Food, Beverage, Tobacco	1537	2249	10677	3861
Textiles	2199	3101	8512	3961
Wood	1307	1971	3308	2356
Paper and printing	221	289	4349	707
Leather	91	847	320	481
Rubber, Petroleum products	250	234	999	1368
Chemicals	2555	770	1896	852
Minerals	6344	2147	6410	2292
Basic metals	28581	49590	55747	47413
	(58%)	(70%)	(45%)	(46%)
Metal products	1610	2577	1795	6473
Machinery	926	404	11221	1199
Electrical equipment	153	207	3363	3760
Other, Repair	3729	5830	11147	6087
Total	50558	70352	124483	101469

Notes: Figures are for the District (urban); the name of the principal steel town is bracketed below the District name; the percentage of workers in basic metal production is bracketed below the figure.
Source: Government of India, *Census of India, 1981*, Orissa, Madhya Pradesh, West Bengal, Bihar, General Economic Tables (Table B12)

Table 5.3
Composition of Family Earners, Income, and Remittances in Industrial
Worker Households in Selected Cities, 1958-9

City	Decadal Popu-lation Growth, 1951-1961	Main Industry	Average Income per Employee in (Rs)	Single Earner House-holds (%)	Other Major Earner	House-holds remitting (%)	Average Level of Remittance (in Rs)
Asansol	78.0	(a)	112.2	89.2	child	34.7	13.0
Jamshedpur	50.4	(a)	158.7	80.2	child	26.3	13.4
Raniganj	16.4	(b)	82.9	85.9	wife	42.5	13.9
Jharia	27.2	(b)	93.4	78.3	wife child,	41.9	12.7
Ahmedabad	37.4	(c)	118.0	77.2	wife	27.6	11.4
Sholapur	21.8	(c)	54.9	53.0	wife	1.8	0.3
Monghyr	20.7	(d)	124.8	81.5	child	13.3	5.3

Key: (a) Iron and steel; (b) Coal-mining; (c) Cotton textile; (d) Cigarettes, Locomotives
Source: Government of India, Ministry of Labour, *Family Living Surveys among Industrial Workers, 1958-9;* Government of India, *Census of India, 1961,* West Bengal, Bihar, Gujarat, Maharashtra, General Population Tables (Table A4)

Figure 5.1
Industrial Employment Distributions
in Durg, Burdwan and Poona

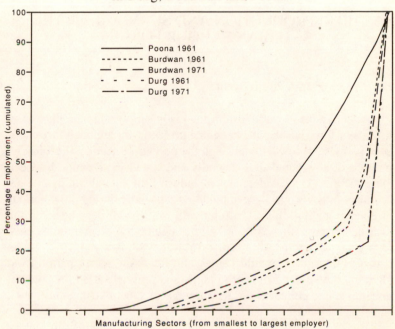

Note: Manufacturing sectors are arranged in order of employment for each District and date.
If each sector employed the same number of people the observations would fall on the diagonal.

Source: From data provided in Government of India, *Census of India, 1961, 1971,*
Economic Tables (Table B4)

Chapter 6

THE REPRODUCTION AND SURVIVAL OF THE URBAN LABOUR FORCE

The Theory of State Involvement

Throughout most of world history the ready supply and reproduction of the labour force has inevitably been a problem; for although there have been periods of growth to outweigh the periods of decline, the long-run increase of both population generally and labour in particular has been close to zero. The development of industry, of capitalism, and of the economy in general have been hindered especially during those periods of demographic decline. The relatively low spatial densities, and the insufficient migration to form large population agglomerations, have retarded trade and industrialization (Marx, 1983; Boserup, 1981). India was arguably no exception. During the relatively recent period from 1890 to 1920, combinations of plague and famine caused the colonial government to comment on the shortage of labour for public works, and even for industry.[1]

But in the early days of India's capitalist development in factory manufacturing, which might be dated approximately to the latter half of the nineteenth century, the state was relatively uninvolved in the direct development of industrial capital, though it facilitated the provision of infrastructure such as railways: indeed it might be argued that the British industrialists themselves, fearing competition, would if necessary have encouraged the Government of India to remain uninvolved (except to ensure 'fair competition' by introducing factory legislation similar to that controlling British industrialists). We find also that the first governmental measures taken to encourage human survival were specifically and overtly directed towards the survival not of the Indian working class but of the British army in India, which was nearly decimated each year by diseases like cholera and dysentery.[2] The result was the first sanitary measures taken to improve the environment in cities like Calcutta, an environment from which the colonial personnel could not be entirely isolated. But it was left to the private industrialists themselves to ensure a sufficient supply, stability, and survival in their own workforces to meet requirements, especially those of skilled labour: in cities such as Jamshedpur (the home of the steel industry) they built pukka housing towards that end.[3]

By the mid-twentieth century, and especially in the wake of the second world war, the major capitalist powers, and in particular the United States, became apprehensive of the need for more widespread military intervention in the east to pre-empt the rise of communism. At the same time, before the world had become obsessed with the fear of massive populations, and before commerce had begun to target consumers from the richer subsectors of those populations only, a large, aspiring, and healthy consumer body was seen as the sine qua non for the development of the postworld war economy in Asia. It had not only been Marxists that predicted capitalism would founder on an inadequate consumer base; the view was orthodoxy, as Keynes' only tract on the population problem, the problem of a *slowly* growing population, indicates (Keynes, 1937). In fact in the 1940s a consultant from the Rockefeller Foundation, which sponsored many of the health programmes, explicitly stated that assistance to improve the health of the populations of Asia would ultimately 'benefit the United States both politically and financially' (quoted in Doyal, 1976, p.275).

It was in this climate that the Indian state, with international assistance, became seriously involved in a programme that dramatically and to some extent unintentionally brought an end to the centuries old problem of ensuring the reproduction of the population. Indeed so successful were the nation-wide health programmes of the 1930s, 40s, and 50s in reducing malaria, cholera, and smallpox that within the space of only twenty years the state had become involved again in population matters. Ironically, this time it was because the population was reproducing *too fast*.

By the 1950s it would be reasonable to argue that the Indian state had become directly involved in the development of indigenous capitalism. Since the formulation of the Bombay plan in the 1940s Indian capitalists had explicitly indicated that they expected the new independent government to take a lead in building parts of the new industrial structure. It became increasingly clear that the state was to take the role of bearer of social costs of the new industrialisation, while private capital reaped many of the benefits (Bagchi, forthcoming). It could also be argued that, in those early days at least, the public sector was expected to promote the emergence of a large industrial working class, appropriately trained and subsequently maintained in healthy working order. Now as we have seen, the crude supply of those workers was at last assured by a natural increase that had leapt from close to zero to approximately two per cent per annum, implying a doubling time of thirty-five years. Why then did the state, especially in the early 1960s, become concerned to check the growth that its health programmes had so successfully brought about?

As we pointed out earlier, the health programmes were not *designed* to increase labour supply, this was an incidental outcome; though some economists indeed have calculated the gain in productivity that would follow from a malaria eradication programme.[4] It is clear from the central government five-year plans that the resultant population increase was expected to be substantial, but not as substantial as it turned out to have been when the 1961 census take was complete. In fact the Second Five Year Plan document of 1956 predicts a 1.2 per cent per annum growth of population between 1951 and 1961; it turned out to be 1.9. The Indian state was much less concerned with the growth of markets than international capital had been: its prime concern was the growth of investment in general and heavy industrialization in particular. At that point in time its attention was also focused on the quality of the labour force: 'Economic growth means not only production but also more, and increasingly more, capacity to produce. In this process the development of human faculties and skills is no less important than the marshalling of material resources' (GOI, Planning Commission, 1956).

It so happens, at about the same time, some western economists and demographers were drawing attention to specific disadvantages of rapid population growth for the growth of the economy. A.J.Coale, who visited India and talked to many academics involved in the planning exercise (including P.C. Mahalanobis) and based his analysis specifically on India, following the parameters relevant to the Second FiveYear Plan, published a study on the subject in 1958 (Coale and Hoover, 1958). It demonstrated the now familiar proposition that high fertility low mortality populations have young age structures: India's under-15 population was about 40 per cent, compared with 20 per cent in western low-fertility populations (and 30 per cent in the Indian population prior to the major mortality decline). Coale argued that most social expenditures on welfare, and especially those on education, were primarily incurred by the need to service the under-15 population. His argument then took a more controversial turn: the economic return on such education, because of the long gestation period that elapsed before the child became a productive worker, was low. It followed that the incremental capital-output ratio (which was one of the linch pins of Indian growth plans in those days) would be high, at least in comparison with that of investment in industrial capacity, which came on stream in about three years. In summary then, increased expenditures on education, due to the youthful nature of the population composition, would reduce national income growth below what it might otherwise have been were the population older. The only way to change the population composition favourably was to reduce fertility, which, now that mortality had been reduced, could be done without bringing the population growth to a halt again.

A second major consideration, also stressed by the economists, was a more straightforward and uncontroversial question of arithmetic. The faster the growth of the population, the faster the growth of the labour force. It follows that a greater proportion of investible resources has to be allocated simply to preserve the capital-labour ratio (capital widening); with a slower growth of labour the same resources could have been used to increase the capital-labour ratio instead (capital deepening).

It would be attractive to believe that members of the Planning Commission and others in government absorbed these arguments, and that they contributed to the greater seriousness with which population control was approached in the early 1960s. This inference would be entirely consistent with the stress on productive industrial investment and the prevailing obsession with the capital-output ratio (which was estimated for some decades ahead in the Second Plan documents). It would also be consistent with another emerging concern of the new industrial state—the quality of the industrial labour force. For to obtain a large increase in qualified manpower, one has to establish high rates of primary and secondary schooling at the base of the educational and demographic pyramid. What was explicitly envisaged was universal elementary education (though not only for the materialistic reasons outlined here). It is also worth remembering that the new industrial labour force was still being recruited actively by the public sector, with literacy as a preferred qualification. The economic requirement for widespread and good quality education was important; but it was an expensive goal.

Despite the logic of this argument, and the forecasts of economists like Coale, what actually happened was rather different. For reasons which have been well-discussed, fertility was slow to respond to the programmes of fertility control, which were not in any case pursued with much vigour or allocated many resources (Soni, 1984); furthermore, demographic momentum kept the age structure almost unchanged until the 1980s. In reply to the undiminished youthful dependency burden thus implied, the state failed to follow the path predicted by Coale.

It effectively abandoned its education targets, rather than cut into its funds devoted to productive investment. The inevitable outcome was a very gradual progress only on the path towards universal literacy; and a steady deterioration in the quality of higher education provided by the state. Resources however continued to pour into industry (though the return to such investment was soon to fall). This may have been a serious mistake, as research has shown that the economic growth-enhancing value of education is considerable (Psacharopoulos and Woodhall, 1985).

It would however be simplistic to attribute the state's interest in slowing population growth to one cause only—the need to avoid diverting resources from quick-yielding industrial capital to long-yielding human capital investment. In fact it is likely that a more pressing need was felt to be emerging by the early 1960s—the ability to feed the growing population; and international aid in the green revolution programme was premised upon a concomitant effort to check population increase, with the focus on rural areas. Besides, we must keep the concern of the state in perspective: neither health nor family planning (except in 1976 during the Emergency) have been accorded a real priority.

As we have said, labour supply in terms of pure numbers had ceased to be a constraint even if its quality had not. The private and public industrial sectors had attempted to alleviate this problem by building captive training institutes. Both sectors took measures to preserve the health of their core labour forces by constructing captive townships and providing health services at work. One result of this rapid increase in the overall supply of workers in the population was a ready supply of labour, after some initial encouragement, to the urban industrial complexes; indeed, as we have shown in an earlier essay, urban natural increase, aided by the demographic momentum in cities that had been subject to heavy in-migration, came to supply raw labour in abundance as it reproduced itself. By the 1980s even the supply of qualified manpower was less of a problem, in that the numbers of educated workers in the major urban labour markets made it less necessary to seek a supply from outside. Besides, private industry had reduced its intake of labour per unit of capital considerably; this was not, of course, to combat a restricted supply, but largely to reduce the potential for industrial unrest and rising wage costs. If the state was now able to target its training and health promoting services more precisely, and the growth of labour demand was diminishing, would it follow from all this that the state would lose interest in overall population control, its reproduction and survival, in urban areas?

The answer, it seems, is that it did not. The reason probably is the rise in the perceived social cost of maintaining a livable urban environment for the Indian bourgeoisie. Although it would be theoretically possible to build infrastructure and housing for an urban domestic core consisting of the industrial labour force, and leave the rest to wallow in squalor, it is *increasingly* difficult to insulate the urban bourgeoisie from the congestion, noise, and smell that such a policy entails (just as before the British had attempted to isolate their personnel in cantonments). On the other hand the urban proletariat has become increasingly articulate and important to urban-based political parties in the last decade or so, especially in the metropolitan cities. The state is under pressure from both sides to increase its expenditure on urban improvements.

Furthermore, both Indian and international capitalism are increasingly eager to focus consumption on a relatively prosperous minority. Economies of scale in marketing and distribution favour urban agglomerations for this purpose. But for such economic systems to function smoothly requires investment in transport facilities, water supply, and housing for a far wider spectrum of the population than just the bourgeoisie or core industrial labour force: the support of transport operators and police are equally important for example. This no doubt explains the recently increasing involvement of the World Bank in financing social infrastructure in the metropolitan urban areas. Many social expenditures of this kind would be less if the growth of urban population were less and its composition less youthful. This is true of urban education and health services (where the demand is high and well-articulated), as well as the highly expensive housing development. With over one half of urban population growth coming from natural increase, not migration, control of reproduction as well as maintenance of health remains on the state's agenda in urban areas.

It is our general contention that the state and private industry both play a more direct role in determining survival than they can do in determining reproduction. For although in both cases the state provides information (both good and bad) that may be acted upon within the household, in the case of survival much is determined in the working environment outside the control of the household.

Reproduction—the Empirical Evidence

The crude birth rate was distinctly lower in the urban areas (30 per thousand) than in the rural areas (38 per thousand) at the start of the 1970s (when the relatively reliable Sample Registration records begin); by the late 1980s the gap had hardly changed (at 27 urban and 34 rural). As we remarked in Chapter 2, the social composition of cities will account for some of this differential, which may have persisted since the turn of the century at least. Recent studies have shown that the difference between the fertility of migrants and that of locally-born women is largely explained by their educational differences (Gotpagar, 1990).

Closely influenced by social factors is the age at first marriage. Marriage age can be a very important determinant of the birth rate; indeed an analysis of fertility change in the 1970s attributed about one third of it to changes in propensity to marry (Jain and Adlakha, 1984). The most detailed information here can be extracted from the 1961 census tables for Bombay, referring to a period rather before the state became much involved in reproduction control. Bombay women married at the age of 19.4 on average. As regards migrants, Table 6.1 shows that working women of rural origin married nearly three years later than non-working women (at 18.2 compared with 15.5); for those of urban origin the difference was nearly six years. To get a better idea of the likely effect of marriage on reproduction it is better to focus on the age groups 15-39, and as an index we can take the two age groups 20-24 and 25-29. Here we can see that among rural women only two per cent remained unmarried by ages 20-24 in the non-working groups, compared with 30 per cent in the working group. In the case of women of urban origin, where the working women are more likely to be of professional status, the similar difference is as great as seven per cent versus 58 per cent. One has of course to remember that the large majority of female migrants are non-working women of rural origin in this city at this date at least. Data from other cities of Maharashtra tend to bear out the argument on social composition (Table 6.2). Sholapur was an old established town, a textile 'monoculture', attracting little migration by the 1950s; not surprisingly a traditional pattern of youthful marriage still prevailed with only 11 per cent of women single by ages 20-24. The contrast with Poona and Thana which were rapidly beginning to modernize is clear.

But let us now turn to some heavy industrial towns of eastern India (Table 6.3). In all the heavy industrial areas of Burdwan District in

West Bengal 90 per cent of women are married by 20-24; this is consistent with our knowledge that they tend not themselves to be working in heavy industry, and that the male migrants to the steel towns soon had their wives join them. The mining towns are even more striking in that 70 per cent of women there were married in the young age group of 15-19. This undoubtedly reflects the strong rural tradition, and the low-caste social composition in these towns; it conflicts with the fact that these women are more likely to be employed. But in mining, employment tends not to bring status or imply career opportunities: it is rather a manifestation of poverty endured as part of one's daily life, and would not be likely to delay a marriage. It is, as it happens, a widespread, if not universal observation that mining families retain high fertility when their industrial colleagues are reducing theirs.[5]

Social composition apart, does an urban environment intrinsically lead to later marriage and lower fertility? It is certainly true that aspirations towards a better material lifestyle are fostered by urban residence, where the examples of middle class consumption standards are demonstrated in close proximity. Later marriage and fewer children enhance the prospects of acquiring these. At the same time, aspirations for one's offspring, and the aspirations of that second generation itself, are heightened in the modern industrial and commercial environment; increasingly this requires one thing in particular: education. The direct and opportunity costs of the latter must be one of the more pressing reasons for private fertility regulation. A further pressure, however, lack of private living space in chawls and shanty towns, also induces smaller families. In other words, what the capitalist state requires seems to coincide fairly well with the requirements of the individual family. Without much effort on the part of the state, there would in any case have been birth control in urban environments.[6]

It is not so easy to trace the progress of marriage and fertility over time as the detailed data of the 1961 census are not replicated in the later censuses. The slow progress of entry into marriage in Bombay continues to contrast with the fast pace in Sholapur through to 1981, to judge from district-level data (Table 6.4); by 1981 just over a third of women aged 20-24 are single in the modern metropolitan and industrial city of Bombay. But marriage age has increased in all the districts selected in Table 6.4.

Within the district of Burdwan we are able to examine in a little greater detail what has been happening in the steel town of Durgapur where the author (with C.R.Malaker) conducted a demographic survey in 1988. The core-sector steel workers were mainly housed in the official township, so a distinction between township and bustee sectors brings out the contrasts between the industrial workers and the rest. In the age group 20-24 as many as 33 per cent of the women were still single from the township stratum (much higher than the average for Burdwan district (23 per cent) and closer to the average for Bombay).[7] The daughters of the original steel workers in the core labour force seem now to be marrying considerably later than did their parents. Their relatively comfortable economic position and educational opportunities have made many of them aspire towards careers of their own and to postpone marriage.

Adequate measures of fertility are not available from census data to indicate the contribution of marriage to fertility differentials or decline. Our survey, however, provides some data on children ever-born by age of mother for the township and bustee strata in Durgapur. There is a very substantial difference between the two strata of younger women (Table 6.5). The core industrial labour force seems to have been reproducing less rapidly as the city matures; this would be an appropriate response to the less rapid expansion of jobs for the second generation in heavy industry, for which education is a prerequisite. For it has to be remembered that the heavy industrial cities have suffered from considerable demographic momentum: by which we mean that the major in-migration of young adult males, followed within five years by young adult women, inevitably resulted in large numbers of children being born. It was a consequence that the industrial planners at the time did not fully appreciate, and we have argued in Chapters 4 and 5 that an appropriately diversified ongoing investment programme would have made some social sense. Encouraging birth control can only reduce, hardly eliminate, the problem, and then not in the short run. It seems that individual families, mostly those living in the steel town proper, aware of these facts, responded rationally within the parameters of their control.[8]

To summarize, so far we have argued that urban living conditions, the industrial employment situation, and aspirations to acquire qualifications, have all encouraged working class households to limit their fertility. Similarly, a combination of career aspirations, and status improved by going out to work, have encouraged women to resist early marriage, with consequent delaying and reducing effects on fertility. These factors have been offset somewhat by age and marital status of first generation migrants with immediate repercussions on the child-bearing propensities in some urban areas. The young age structure of urban populations demanding education, housing and social services of the state, and the supply of labour to industry in excess of the need to keep wages down, have persuaded the capitalist state to take a role in reducing the rate of reproduction also.

Survival—the Empirical Evidence

It is not really very helpful to examine the crude death rate data from the Sample Registration System. At the relatively low levels of mortality that characterize India's urban population the crude death rate is strongly influenced by age composition: as fertility falls and the urban population gets an older structure, the death rate is bound to rise simply because older people are more prone to death. Hence while the rural death rate has fallen from 17 per thousand in the early 1970s to twelve per thousand in the late 1980s, the urban rate sustained a smaller absolute fall from ten to eight. As with our analysis of reproduction, we will focus here on policies and behaviour that make for mortality differences between and within cities.

The survival of the labour force becomes of immediate interest to the capitalist state if it has made a valuable investment in human capital. This is most easily appreciated at the level of the enterprise; and it is here that intervention is most readily targeted. Amenities provided at the workplace include such health promoting facilities as latrines, canteens, and medical services. These were recorded from the responses to the Labour Bureau surveys of the industrial workforce, carried out in selected cities in the late 1950s (GOI, Ministry of Labour, 1968). We have categorized a sub-sample of eighteen of those cities according to the major industry present in three groups: heavy industry, textile industry, and mining (Table 6.6). A score of one is allotted for each of the above-mentioned facilities that is provided, and the percentage of the maximum possible coverage is calculated for the cities in each group. It can be seen that the provision of such facilities is best in the heavy industrial towns, next best in the lighter textile towns, and considerably worse in the mining towns. This corresponds well with our discussion on the need of the capitalists to protect their most valuable workforce. Replacement, including training, of the core labourers in the engineering or steel-making industries is costly; in mining it is less so. Of course to complete the cycle of labour reproduction requires the provision of wage goods and basic services (not just health services) as well as an appropriate level of human reproduction: this in turn requires a labour force to provide the wage goods. But only in the case of extreme labour shortage would one expect the industrialist to take measures, or require the state to take measures, to provide attractive amenities for this sector of the workforce. And it is only occasionally in the past that such sectors of the workforce have been able to organise sufficiently to demand effectively such facilities for themselves.

In order to assess the outcome of such differential approaches for survival, we need ideally to obtain data that distinguish the various factors that may contribute; such sophistication is not possible in practice. For cities that have healthy workplace environments for some of the working class, in terms of facility provision at least, may also have well-built and well-serviced residential townships for the same workforce and its families. By contrast, miners work in the worst conditions and live in those little better. For what it was worth, we attempted to construct an index of male mortality for 26 urban districts of varying industrial character in the eastern region of India; the index we used was widowhood among women aged 35-49 (Table 6.7). The evidence is far from definitive, but it suggests that there is lighter mortality where there is heavier industry, which is what we would expect from our discussion above.

Living and working environments interact with wage levels to affect the health of the whole household; this is important to consider when accounting for nutritional status (an outcome of both food intake and absorption, the latter determined by health status). A similar picture to that presented above is suggested if we review some of the data available on nutritional status of the urban population. Nation-wide studies carried out in the late 1970s show the average calorie intake to be higher in industrial worker families than among slum dwellers in general (Jaya Rao, 1985). The author's own study in Durgapur (with C.R.Malaker) found that child mortality was higher among working class households living in bustee settlements than among those living in the steel township, who were all predominantly engaged in employment in the steel industry (Bapat, Crook and Malaker, 1989). The study further showed that both occupation and environment could independently account for mortality differentials. The gap in survival between the various sectors was substantial: the infant mortality implied by the child survival rates was 46 deaths per thousand in the township versus 79 in the bustees; and 59 for children of the core sector labour force versus 94 for the rest. It need hardly be stressed that in a new steel town of this kind the environment is almost wholly 'planned' by the state industry.

Unfortunately studies and data are lacking so far with which to illustrate the effects of state remedial intervention, but those that exist suggest that better-serviced slum environments enjoy better levels of human health and nutrition (for a study in Poona see Bapat and Crook, 1989); international capital in the guise of the World Bank has assisted programmes in Indian metropolitan cities, starting with Calcutta in the 1970s, followed by Madras and more recently Bombay. Its investment has also started to penetrate the heavy engineering towns. Last on the list will be the smaller light industrial towns, the old stagnating industrial cities, the mining villages; these have the least articulate middle class to complain of environmental nuisance, the least potential for growing consumer markets to cultivate, the least organized working class to protect itself from the materialism of the capitalist state. They certainly do not have the lowest mortality. If a desire for equity had been the driving force, they would have commanded attention long ago.

In summary we would suggest that a significant factor in determining the survival of the population is its closeness to the spearhead of modern industry. This reflects its perceived importance to public and private capital, and the investment embodied in labour itself.

A Postscript

This essay has illustrated a materialistic view of human reproduction and survival in the context of an industrializing economy. The writer does not condone this view. Nor does he regard it as a universal truth. In a just society prolongation of human survival should be an end in itself, regardless of the economic value of the person to the state. Given reasonably healthy economic, social, and familial circumstances, people of all ages enjoy living; children, long before they become productive, give joy to themselves and to their parents. They create welfare, or consumer services if you like, simply by being. Presumably pre-productive, half-finished, steel mills give little joy to their capitalist owners, and are incapable of joy themselves. Therefore, to compare the two by the same yardstick as economists are prone to do, is wrong. For similar reasons the state has no moral business in controlling reproduction in order to raise per capita incomes; though it may be entitled to intervene to some extent if it is already subsidizing the costs of children to their parents who thus reap artificially elevated benefits from childbearing.

Nor in practice does the state behave in a purely materialistic way. Cross-currents of politics temper the tide. Political parties respond to popular demands for sanitary reform. Populist politics divert attention from the roots of poverty by responding to clamours for birth control. Science warns that unlimited population growth per se must eventually become a problem. Religions preach tolerance, and great men in high places have genuine concern for humanity. The state in other words is a plurality. Nevertheless, the tide beneath the cross currents is fundamental. In the context of a growing industrial economy, the capitalist state's involvement in reproduction and survival deserves to be considered in this light.

NOTES

[1] It is of course possible for there to be a co-existence of a relative surplus population in a static sense and a shortage in a dynamics sense: industrial capitalism prefers a pool of unemployed to keep wages down, but also requires a steady growth in employment (and unemployment) to facilitate the accumulation process. The bottle-neck can be broken by substitution of capital for labour, but this process cannot proceed faster than is permitted by the rate of accumulation.

[2] Reference here may be made to the first Bengal Sanitary Report. Mortality rates of between 50 and 60 per thousand in the army are quoted (GOI, Department of Health, 1865).

[3] As M.D.Morris points out (Morris, 1965), it is often difficult to distinguish between shortages of labour generally and shortages of skilled labour. He is inclined to the view that, even during the high-mortality years at the beginning of the century, absolute shortages of labour were rare. The even more difficult 'counter-factual' problem is to know whether there would have been a more rapid industrialization if there had been a more rapid growth in the raw labour supply.

[4] Cost benefit analyses of malaria eradication were undertaken for the Philippines, for example. Rather cruder calculations were made of the effect of eradicating shistisomiasis from parts of China in terms of increased output (Hillier and Jewell, 1983, p.165).

[5] This has been observed in the mining towns of Wales during the latter phase of the industrial revolution in nineteenth-century Britain and Germany, also in the United States (Weber, 1963).

[6] One of the clearest demonstrations of this fact comes from Shanghai, China, where living space was much reduced by the 1970s. Well before the state stepped in to promote single-child families, a substantial proportion of households in that city (though not elsewhere in China) had already adopted this norm of severe birth limitation.

[7] These data are of course subject to sampling error: the sample consisted of 1500 households altogether in both township and bustee strata.

[8] The only crude measure of fertility that can be obtained from all the censuses is the child-to-women ratio. The usual age groups to take are 0-4 for children and 15-49 for women. It has the disadvantage that if child survival improves, the ratio increases, giving a false impression of fertility

increase. For what it is worth, however, we calculated child-to-women ratios for urban districts of West Bengal. In Burdwan there was very little change between 1961 and 1971 (a decline of 3.3 per cent), but between 1971 and 1981 a decline of 34.1 was recorded. In most other districts there was less contrast between the two decades.

Table 6.1
Marital Status of Migrants to Bombay, 1961

Age	Percentage of Female Migrants Single			
	Working		Non-Working	
	Rural Origin	Urban Origin	Rural Origin	Urban Origin
10-14	88	96	90	89
15-19	41	73	19	36
20-24	30	58	2	7
25-29	7	26	2	10
30-34	4	15	1	3
35-44	2	7	1	2
45-59	2	5	1	1
SMAM	18.2	23.0	15.5	17.2

Note: SMAM is the 'Singulate Mean Age at Marriage', i.e. the average age at which the single marry for the first time.
Source: Calculated from Government of India, *Census of India, 1961*, Maharashtra, Special Migration Tables for Greater Bombay (Table 6).

Table 6.2
Marital Status in Three Contrasting Cities of Maharashtra, 1961

	Percentage of Women still Single by Age Group		
	15-19	20-24	25-29
Poona	55	19	5
Thana	60	19	6
Sholapur	35	11	1

Source: Calculated from Government of India, *Census of India, 1961*, Maharashtra, Cities of Maharashtra (Table C2).

Table 6.3
Marital Status in Contrasting Urban Localities of Burdwan District,
1961

	Percentage of Women still Single by Age Group		
	15-19	20-24	25-29
Asansol (heavy industry)	40	10	4
Hirapur (Burnpur steel)	50	11	2
Durgapur (steel)	41	9	6
Chittaranjan(loco manufacture)	54	8	3
Kulti (coal mining)	23	6	8
Raniganj (coal mining)	29	8	4
Ondal (coal mining)	27	7	5
Burdwan (district HQ)	43	19	11

Note: Localities are the urban sectors of the 'Police Station' areas; economic
character of the locality indicated in brackets.
Source: Calculated from Government of West Bengal, *Census of India, 1961*,
Burdwan District Census Handbook, Volume 2 (Table C2).

Table 6.4
Marital Status in Selected Districts (Urban) of Maharashtra in
1961, 1971, 1981

	Percentage of Women Single by Age Group					
	20-24			25-29		
	1961	1871	1981	1961	1971	1981
Bombay	22	30	34	8	8	11
Thana	23	22	26	4	5	7
Poona	17	20	24	5	5	7
Sholapur	10	14	17	2	3	4

Source: Calculated from Government of India, *Census of India, 1961, 1971,
1981*, Maharashtra, Social and Cultural Tables (Table C2).

Table 6.5
Fertility of Women in Township and Bustee Localities of Durgapur,
1988

Age Group	Children Ever-Born		
	Township	Bustee	Combined
15-19	0.0	0.3	0.1
20-24	0.4	1.7	1.2
25-29	1.6	2.6	2.3
30-34	2.3	3.2	2.8
35-39	3.0	3.9	3.3
40-44	3.2	4.6	3.5
45-49	3.9	4.7	4.1

Source: Survey reported in Crook and Malaker (1988).

Table 6.6
Welfare Facilities provided at Place of Work in Towns of different
Industrial Character

Character of Town	Number of Towns in Sample	Percentage of Workers Reporting Facilities
Mining	8	63
Cotton Textile	6	85
Heavy Industrial	4	91

Note: The percentages are the unweighted average of the percentage of workers
reporting provision of the following:
1) latrines, 2) canteens, 3) medical facilities.
Source: Calculated from Government of India, Ministry of Labour, Labour
Bureau, *Family Living Surveys among Industrial Workers, 1958-9*.

Table 6.7
Incidence of Widowhood in Urban Areas of Different Industrial
Character

Character of District	Number of Districts (urban)	Women Widowed (%)
Light Industry	12	22
Mixed Industry	9	20
Heavy Industry	5	19

Note: Percentages are for ever-married women in the age-groups 35-39, 40-44,
45-49, an unweighted average being taken.
Source: Calculated from Government of India, *Census of India, 1961*, Bihar,
Orissa, West Bengal, Madhya Pradesh, Social and Cultural Tables(TableC2).

Chapter 7

URBAN ECONOMIC MANAGEMENT: THE QUESTIONS OF SIZE AND CONGESTION

The Economic Advantage of Size

As the cities of India grow to vast sizes two assumptions are readily made.[1] The first is that, from an economist's viewpoint many of them have become too large. The second is that an optimal policy to meet this complaint involves the clearing away of make-shift dwellings that have encroached on public land, and the removal of traders and hawkers from the streets; this, it is said, will discourage further migration, and eliminate the increase in slums and congestion that characterizes these overgrown cities.

This essay and the essay on housing (Chapter 8) will challenge both of these assumptions simultaneously. In so doing it will review a number of urban policy problems to which better economic solutions than those currently in practice are readily available, and the implementation of which would be to the advantage of *all classes* in society.

Urban occupations encompass a range of labour productivities from the most to the least productive members of the workforce. The rapid growth of a city is an indicator (though not proof) that the productivity of its workforce is also increasing. Bombay and Bangalore for instance would *not* be growing were they not prosperous. Indeed there are several large cities that are not growing faster than their natural increase alone (i.e. there is no net in-migration) (Table 7.1).[2] Those people who are forced into a city owing to imminent destitution in rural areas come to those cities where they can be assured at least of some transferred income from the more productive among the poor. Without the high productivity there would not be the high migration even of the destitute, as can be seen by the low rates of migration into smaller, less prosperous, but nearer towns in rural areas of poverty or at times of drought. And if the poor wish to share their income among themselves, that is up to them; just as the rich share their incomes among themselves to help educate younger family members for instance.

What is sometimes forgotten is the dependence of the more productive workers and the managerial classes on the less productive services sector. If that sector is excluded from the city, or its supply price raised by shifting its residential quarters to the outskirts, the prices of services must increase. Evidence for this is clear in the reported rise in the cost of service sector workers in New Bombay, for example, consequent upon the shifting of the newly developing squatter colonies. Similarly in the Fort area, the central business district of Bombay, simple services such as shoe-shining became hard to come by, and the bargains struck tougher, after the Corporation had become more resolute in its restriction of traders. In other Asian cities where extreme restrictions on migration were enforced, as in Shanghai for example, acute shortages of domestic service labour were reported, a problem enhanced by the ageing population structure consequent on the sharp reductions in natural increase.[3] These are all anecdotal examples. But they make a straightforward and universal point: any *indiscriminate* policy of exclusion excludes the more productive and the less productive alike within any occupation or economic class. The result is a rise in price in the services and traders' products, suffered by those who presumably represent middle class interests, in trying to restrict supposedly congesting activity. That is to say, it is possible this way to cut off one's nose to spite one's face.

A word here is perhaps necessary as to why large cities should harbour a particularly productive labour force. The fact is related to economies of scale in industry. This does not necessarily mean the economies of scale enjoyed by having large factory plant (on which we elaborated in an earlier essay, Chapter 4). The agglomeration of commercial and industrial undertakings located in close proximity to one another enhances the efficiency of communications especially in societies where telecommunications are relatively costly (i.e. in capital-scarce economies, like India's). If average incomes were to reflect these advantages and enhanced productivities, they would rise as city size increased; this phenomenon has been well documented in Japanese cities (Mera, 1973) and to some extent in India also (Shukla, 1988). It is partly reflected in wage rates in Bombay and other metropolitan cities, and in the high per capita municipal revenues in the class I Indian cities (i.e. those of at least 100,000 people) in contrast to those in the smaller size classes (Datta, 1968) (Table 7.2). If this were the last word on the subject the message would be: the larger the city the better. But an important qualification has to be made, to be discussed later.

Urban Problems and Appropriate Policy Responses

The removal of slums, especially squatter settlements, is often sought for hygienic and aesthetic reasons. A preferred policy nowadays is to remove the population to a site typically further from the centre of employment (which has the unfortunate side-effect of raising the price of labour as we mentioned above; Bapat, 1981). Sometimes no alternative site is provided, or the new site is poorly drained or serviced, one of the objects being to discourage the population increase altogether. This policy implies incorrect diagnosis of the problem. To the neoclassical economist the problem is a straightforward one of incorrect pricing of public services. Since water, sewerage, and drainage services are not provided on the original squatter sites (nor indeed on many sites where hutment dwellers have legal tenure), the occupants use existing public land or water as if it were a free good: any such use rapidly becomes over-use. Water sources run dry and drainage becomes clogged. The correct solution is not to exclude the population from facilities, but to provide those facilities and charge for their use. This is most equitably done on a community basis (i.e. per hutment colony, or by sub-dividing colonies) so that the community can decide on the per household charge, subsidizing its poorer members from its richer members if it sees fit. This is in recognition of the fact that some migrants may have been forced to come to the city to escape death from famine or flood; and some urban-born households, with no rural base, may be forced into destitution through the death or incapacitating illness of the principal breadwinner. In such cases households could not bear the charge of the social services they use, but to drive them out of the city would be quite unethical. It should be emphasized that such community organization is to be found in numerous (but not all) slum settlements in the large cities, often originally formed for the collection of temple funds, or contributions to a milk service, or for the creation of nursery schools. It is a hallmark of economic development that such organizations get formed and are allowed to operate. Ignoring this possibility and either seeking to remove, or failing to service, slum settlements altogether is by contrast a policy of backwardness. It is inefficient; in fact everyone loses out.[4]

The problem with large cities is that the unit costs of providing infrastructure for domestic amenity and common industrial use often rise with population size and density once a particular threshold has been reached. Water sources further afield have to be tapped, sewerage and drainage need more sophisticated treatment (and in the older cities disruptive rebuilding is implied) to ensure the maintenance of adequate standards: all these factors will reverse the trend of falling unit infrastructural costs at some point depending upon the city and its location. That is no problem if its citizens are prepared to pay the rising costs, as they often will be if their productivities (and concomitantly incomes) are higher in the larger cities (as we have suggested above); but the city management cannot tell whether there is a latent demand for these services at their economic price unless it first makes the provision and implements the charge The city continues to be of optimal economic size *however large it becomes*, so long as the increasing population is ready and able to find the wherewithal to pay the appropriate charges. When it cannot, in-migration will cease, or, if that does not follow, migrants or locally-born who fail to pay the charges can then, and only then, be legitimately excluded. What is wrong, is to preclude the opportunity for hutment settlers to purchase services by simply refusing to provide them. That is bad economics, and the *cause* of the lack of hygiene and the disamenity.

A problem that arises, of course, is how to set the tariff for infrastructural services. If unit costs are rising then it is possible to charge the marginal social cost for the use of a service and to cover the intra-marginal costs by so doing. Generally speaking, both locally-born and migrant should pay the marginal cost, for it is both natural increase as well as migration that have caused the costs to rise (and in another essay, Chapter 3, we argue that demographically speaking the two are conceptually difficult to distinguish). In cases of towns or cities of rapid growth, migrants will indeed be contributing disproportionately to costs, and the locally-born may feel it is unfair for themselves to have to pay the resultant charges. However, rapid growth towns are often of small or medium size (under 100,000 population) to begin with, in which case average costs are likely to be falling not rising (the question of initial capital cost being a separate one). In this case nobody will mind if all are charged at the same rate, which has to be high enough to cover average costs, but is likely to *fall*, not rise, as more migrants enter. A qualification to what we have said is that, as we have mentioned above, there will always be cases where in the name of reasonable justice some groups of urban citizens may be charged less and others more; whether this can be income-related, and who will make the assessment, may depend on the circumstances of the individual city. The latter qualification should not lead to excessive city size as long as potential in-migrants generally believe they will be charged for the services used. The truly destitute, who may be relieved of such charges, are not usually in a position to make a *choice* in any case.

It is not central to our purpose here to repeat or develop the arguments relating to optimal tax or tariff structures for local government. However, we should observe that local property taxes, sales taxes, or taxes on the import of goods into the locality (octroi) are all ways of raising revenue for expenditure on public services in India; but none of these is efficient in the sense that each fails to relate the public's use of a public resource to the cost of supplying it. Hence for example citizens could freely enjoy water from the mains supply but be ineligible to pay any local tax; or even if they did pay a tax indirectly (by consuming goods that were subject to octroi), they would not be discouraged from the wasteful use of *water* if the octroi were increased. User charges, on the other hand, can be efficient in this sense. They may, however, be costly or even impossible to administer in some cases: you cannot easily charge (by metering) for the volume use of drainage capacity; and they may be difficult, if not impossible, to make equitable.

A similar point needs to be made about congestion of market areas by street traders and hawkers. We mentioned above that complete exclusion of these simply raises the price of their wares to everyone's disadvantage. The problem should be identified as lying in the free use of public space. A rental charge in the congested areas would cause the traders to disperse and spread out without abandoning the central business district altogether. Quite a small charge would probably effect a substantial dispersal. To be sure, the rent gets passed on in final prices, but the impact is smaller than that from blanket exclusion policies, which drive up prices substantially in the central business area, and tend to shift the congestion to elsewhere in the city. Cities have not generally grown too large as one might be tempted to conclude from these diseconomies of congestion (market places are congested in small towns also); it is simply that they are badly organized. To take an example, Calcutta's central business district around Dalhousie Square has a dearth of cafes. If street food sale were prohibited there would be overcrowding and higher prices in the few cafes that there are nearby. The licensing and the collection of rental charges from the street stalls would improve their hygiene (this being a condition of licence), and reduce their numbers in streets where congestion is severe. Those excluded would set up shop in neighbouring areas and catch the office worker on his way to the congested area. The total number of stalls would be little reduced and the prices not greatly affected (just enough to pay for the reduced congestion that nearly everyone would like). It is true that not all stallholders would gain equally from this strategy: some would lose their most wealthy customers but would enjoy a better environment in which to operate. Questions of equity will always arise from solutions of this kind. Theoretically the losers could be compensated by being given free access to a good water supply by the municipal authority, this being paid for from the rents charged in the central business area.[5]

It would be wrong, however, to think that congestion only arises in the domestic sector. Industrial activities lead to the over-use of both water and air resources. In India there has been an attempt to restrict large industrial plant from locating in large cities. There can be no rationale for such a policy in terms of pollution control. Why should polluting industries be allowed to pollute the air and water enjoyed by smaller towns and cities? It is not the existing level of pollution that matters but the act and effect of pollution itself. Each creator of pollution should be charged the cost of his nuisance: if that cost is higher in larger urban areas than in the smaller towns (perhaps because of existing levels of pollution or the inability of the physical environment to absorb it), then the charge should reflect that fact. There should be a differential tariff. It may seem an impossibility to estimate the charge to set, but in practice this is not a crucial problem. For the result of setting too high a charge nowadays is not usually to force valuable business out of town: given the state of technology and the high cost of some of the raw materials that are leaked into the atmosphere, the offender usually finds it cheaper to install anti-pollution devices and continue to enjoy the economies of agglomeration that keep his other costs lower in the big cities.

In those cases where charging for the use of all domestic and commercial space and facilities (as outlined above) would lead to a large exodus of urban population and capital, the economist might be tempted to argue that the city in question has become too large. In fact the notion of size here is not particularly helpful. The problem is one of congestion, and that can occur in towns of any size, depending on (among other things) their physical configuration and local ecology.

Appropriate Policy in City Systems

There is economic virtue in consistency. If one city adopts the correct public pricing of amenities, so must all the rest, and especially those with close economic linkages with one another; often (though not necessarily) this will mean close proximity. If new-town areas, for example, fail to levy charges, they will receive a larger residential population than they should, some of whom will commute to work in a nearby city (where it would be more expensive to live because the charges are levied). The result will be undesirable congestion of linking roads and transport services; and the new-town will suffer congestion of living space and facilities and may gain nothing (through having inadequate taxation) from the productivity of the resident workforce. If on the other hand a new-town simply prohibits the development of hutment settlements, or refuses to provide adequate amenities, then the nearby cities will receive a larger population than they should in terms of a regional optimum. In this case potential residential space goes underutilized in the new-town, and again congestion of linking roads may occur.[6] Uniformity is crucial; and there is an economics rule that says if you cannot control the pricing policy of your competitors, you may have to modify your own. This would become necessary if, for example, the city that levied the correct rents and charges for facilities etc. was unable to attract new industries to locate there owing to the competition of cheaper labour available in the new town where domestic charges were not being made (and wages could therefore be lower). In such circumstances it might be desirable (from the city's point of view) to allow the growth of rent-free squatting areas. But the best solution is for all to be persuaded to follow the optimal pricing system.[7]

Uniformity may ensure efficiency; it does not always ensure equity. *Within* cities we have suggested that this problem at least be left to local communities themselves; there is no virtue in the authorities restricting the activity of its least productive civic members: such restrictions discriminate against those genuinely hard-working (or those seeking to work) among the urban poor. But *between* cities there would be some virtue in a public redistribution of resources used for domestic infrastructure. We have stated that the size of growing cities should be determined by a combination of their productive capacity and the willingness to pay for the costs of growth. But there is a group of slowly growing cities (with no net in-migration) where the market determination of their size and demographic composition has resulted in an inadequate social infrastructure to provide the minimum respectable services for the rich and poor alike (Table 7.1). The rich or middle classes suffer simply because some minimum threshold municipal revenue is needed before any infrastructure can be laid down or adequately upgraded or maintained. Older cities, both large and small, have often lost their economic base as the economy restructures. Decay sets in, the urban public fabric crumbles, streets fall into disrepair.

There is often an exacerbating demographic factor. Cities subject to substantial out-migration are saddled with aged populations, unable to make an economic contribution and hence unable to raise the level of the municipal tax base. Figure 7.1 shows some striking differences at the top of the demographic 'pyramid'.

Equity would not tolerate these cities simply to be unwound and their populations dispersed or allowed to decay and die in situ. A case could be made for a greater redistribution of resources among towns and cities than currently occurs from national and state expenditures. It is still the case that urban services are substantially provided for from municipal revenues. Seen in that light the huge difference in per capita expenditure between, say, Bombay and Sholapur, not to mention the smaller towns in Maharashtra, is not entirely desirable (Table 7.3). Some minimum level of social service should be ensured for all.

Postscript on the Real World

Although this essay has tried to remain moderately realistic, it is essentially an abstraction from the fussy detail of reality: it depicts a semi-ideal world towards which we might gradually move. The fact is that in India, as elsewhere, we are somewhat distant even from the local optimum. In particular, the rural economy is in no way competitive. The rural powerful control the lives of those born into assetless households, and the weather delivers exogenous shocks that the latter are least able to survive. Hence rural labour is squeezed off the land. In such a situation cities have to be regarded in part as refugee camps. Efficient cost-pricing rules to discourage migration should be unthinkable. Here there is a dilemma: the only instruments we have to ensure efficiency will, if used indiscriminately, also ensure inequity, for which compensation can rarely follow in a land where inequity is often a matter of life or death. In practice these instruments are used with considerable discrimination, as powerful interest groups compete for their use: that outcome does not ensure reasonable justice either. Hence this essay is not meant to be as prescriptive as might appear at first sight. Rather it is intended to prompt some serious and critical reflection on the basic assumptions outlined in its opening words.

NOTES

[1] For a brief overview of the pattern of urban growth see Crook and Dyson, 1982; for a more detailed account see Mohan and Pant, 1982.

[2] Cities also acquire larger populations by redrawing their boundaries to encompass the growth of suburbs. On this see Skeldon, 1986.

[3] This observation was made to the author during a visit to Shanghai in 1982. It has been further pointed out by a referee that it is not only government policy that may keep the price of services high at the city centre. Labour is cartelized in many cases, and new entrants fear for their safety if they undercut the agreed rate. Congestion is therefore transferred to city suburbs (where the clientele is less desirable). The solution is the same as that suggested below: if the state maintains a strict control over space it can decrease *or increase* the service employment on the streets in a particular location, so as to achieve a balance between price of services and cost of congestion. Such intervention will help break the cartelization of the labour market also.

[4] The author was conscious of a range of such organizations in the city of Poona where he worked. Sometimes they were insubstantial and often they were fraught with political conflict, but rarely were they non-existent. More significantly, in conversation with the supposed representatives of these groups the author always encountered an interest in environmental cleanliness and sanitary facilities; there was also concrete evidence that such an interest had in many cases been translated into pressurizing the local authority. The author also interviewed authorities and planners in other cities where such facilities were not provided; they expressed a general scepticism on the feasibility of such schemes, especially regarding the question of maintenance. It is an almost universal experience that social schemes are regarded as impossible where they have not been tried (even if all social groups or classes stand to gain). To quote from the Kerkar report (Kerkar, 1981), 'it did not take long for the (High Power Steering) Group to realize that a "communication gap" exists between those who live in squalor and those in authority. It is this gap that has to be narrowed for the effective implementation of any recommendation and to avoid indifference, litigation, and social protest.'

[5] This would appear to run counter to our earlier argument on water charges. Economists would argue that a correct compensation should be a lump sum. Sometimes a more pragmatic solution is called for, however.

[6] Disputes between the municipal authorities of the cities of Poona and Pimpri (part of Pimpri-Chinchward New Town) in Maharashtra arose during the rapid expansion of the latter in the 1970s and 1980s for reasons of this kind.

[7] There is an academic literature on the optimal taxation (rather than user charges) to ensure allocative efficiency of populations between cities. Generally similar conclusions are reached to those above regarding, for instance, the need to levy differential taxes in line with the differential costs of service provision. This will prevent undesirable migration between cities: those which have become more costly to service, perhaps because

the terrain makes it increasingly difficult to extend the water lines, will levy higher taxes, and hence tend to deflect further migrants to other less costly towns. But taxation does not ensure optimal use of infrastructural facilities by migrants once they have migrated into a city, for reasons discussed above also. Migrants may, for example, move into a city with a low local poll tax in line with low public expenditure on infrastucture and low demand for its facilities, like water supply. Once arrived there they may consume twice as much water per head as the local inhabitants, causing shortages and queueing. Again, differential user charges that move in line with differential congestion in different cities would be more likely to limit the migration (for the correct reasons), or to change water consumption habits of the migrants (equally desirable). See Wildasin 1986.

Table 7.1
Selected Cities of Low Growth Rates

	Average Annual Percentage Growth		
	1961-71	1971-81	1981-91
MAHARASHTRA			
Sholapur	1.6	2.6	1.9
Ahmadnagar	2.2	2.0	2.0
BIHAR			
Bhagalpur	1.8	2.5	1.5
Monghyr	1.3	2.3	1.5
WEST BENGAL			
Bankura	2.3	1.8	1.9
Baharanpur	2.4	2.4	2.1

Source: Calculated from Government of India, *Census of India, 1981*, General Population Tables (Table A4), and *Census of India, 1991*, Provisional Population Totals: rural-urban distribution.

Table 7.2
Index of Municipal Per Capita Expenditure in Selected States According to Size of Town, 1971

State	Size Class of Town					
	I	II	III	IV	V	VI
Uttar Pradesh	100 (31.3)	81	75	76	34	55
Madhya Pradesh	100 (27.6)	92	84	67	43	19
Bihar	100 (13.0)	97	83	44	129	94
Karnataka	100 (33.4)	86	57	49	48	60
Tamil Nadu	100 (47.3)	55	49	35	18	52

Notes: Per capita expenditure in rupees in class I cities is indicated in brackets; the rest of the data are indices with the Class I expenditure set equal to 100. The data refer to years close to the 1971 census.
Source: Calculated from Government of India, *Census of India, 1971*, Uttar Pradesh, Madhya Pradesh, Bihar, Karnataka, Tamil Nadu, Town Directories.

Table 7.3
Per Capita Revenue from Municipal Taxes and Rates for Selected
Towns of Maharashtra, 1971

	Taxes + Rates in Rupees Per Capita
Bombay	71.7
Pimpri-Chinchwad	52.8
Poona	47.0
Thana	40.5
Sholapur	33.2

Note: Figures are for years close to the 1971 census.
Source: Calculated from Government of India, *Census of India, 1971*, Maharashtra, Town Directory.

Figure 7.1
Contrasting Urban Age Distributions, 1971

Slowly Growing Cities

MONGHYR 1971
(population: 102,474
decadal increase: 14%)

SHOLAPUR 1971
(population: 398,361;
decadal increase: 18%)

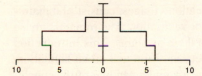

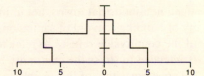

Fast Growing Cities

THANA 1971
(population: 170,675;
decadal increase: 69%)

BOKARO 1971
(population 107,159;
new town)

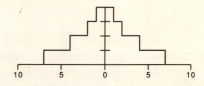

Note: Horizontal axes give percentage distribution standardized for five-year age groups; four broad age groups are calibrated on the vertical axes: 0-14, 15-34, 35-59, and over 60. Males are on the left, females on the right. These diagrams are for illustration only and are not drawn strictly to scale.

Source: From data provided in Government of India, *Census of India, 1961, 1971*. General Economic Tables (Table B2).

Chapter 8

THE ECONOMIC SOLUTION TO THE HOUSING DUALITY

Two popular images are conjured up by the industrial town, one of columns of factory smoke, the other of the residential slum. The latter is typified in the imagination by vast areas of straggling shanty town. That picture is incomplete if not actually erroneous. Slum localities when defined as 'areas where buildings are unfit for human habitation by reason of dilapidation, overcrowding, faulty arrangement and design of buildings, narrowness of streets, lack of ventilation and light' (GOI, Registrar General, 1981a, p. 29) would include large areas of inner city tenement blocks such as the chawls of Bombay, and, in the metropolitan cities, long strings of pavement dwellings. Furthermore, most small non-industrial towns, not to mention villages, consist of a majority of buildings devoid of sanitation or ventilation. The romantic image of the countryside refuses to refer to these as slums. This introduction is intended to stress the continuity between rural and urban habitation, for an appreciation of this enables a ready understanding of the conclusions we will reach.

Housing Demand

The migrant who is attracted or recruited into the city when the expansion of industrial enterprise occurs may earn a wage that is 50 to 100 per cent above that in the rural area from which he or she comes. That gives the migrant a capacity to rent, or purchase on loan, a house costing not more than twice that of a rural dwelling. We shall show below that (once one has allowed for land development costs) such a house will turn out to be even less substantial in terms of construction materials and space (though it may be better serviced with facilities) than that from which the migrant came. Industrialists have long been aware of these facts, and also appreciate that if they abandon the housing question to market forces they will either get a waged labour force that they cannot afford, or else one that is here today and gone tomorrow. For the kind of housing that could be afforded by labourers whose wages can only reflect the productivity of the industry, will be run up hurriedly without sanitation or water supply, at the instigation of private landlords eager to extract maximum rent for minimum service. The result is a dissatisfied labour force, often unprepared to inflict similar discomforts on wives and children, and hence regularly returning home. Additionally it may become a labour force afflicted with sickness and reduced by mortality when tuberculosis sets in at the age of 40 or so. This does not seem like a good return on human capital which the industrial employer will have enhanced through on-the-job (if no more formal) training. Hence he is prepared to contribute an increment towards housing from his industrial profits, and may indeed build the housing himself and rent it out. This practice can be illustrated from the time of the chawls established by some of the cotton textile mill owners of the nineteenth century to the working class bungalows and flats in the steel townships of the twentieth. Furthermore, as we have shown from the evidence of the Labour Bureau Surveys in the essay on health and reproduction (Chapter 6), housing with adequate facilities is more likely to be provided nowadays by those industries that regard their core labour force as scarce: for example, heavy engineering. Employers also build housing for mineworkers, but it is ill provided with facilities. In the textile towns employer ownership is rarer (private rented accommodation being more typical) (Table 8.1).

If investment levels in housing are not to be allowed to rise on a permanent basis, then housing costs must keep in line with urban worker productivities. As we indicated above that the rural-urban income differential reaches 100 per cent for industrial workers, this means that dwelling costs (say in the form of rents) borne by the industrial worker should not exceed rural dwelling costs (which usually means maintenance costs) by more than 100 per cent. That is to say, if we take the value of pukka rural housing to be between Rs. 8,000 and Rs. 16,000 for a landless labourer or small farmer (an estimate based on rural compensation evaluations), then the new urban investment should not exceed Rs. 32,000 per dwelling for the core-sector industrial worker (i.e. about twice his industrial income).[1] For the remaining 75 per cent of the working class population who are not core-sector workers, housing costs must approximate those in rural areas experienced by the lower income groups (i.e. should not exceed an average of Rs. 8,000, with 4,000 being within the means of the underemployed landless labourer living in a flimsy dwelling, whose urban income, say as a casual labourer, would be little more than his rural income was). One simple but striking observation emerges from that requirement: urban housing has to adopt rural construction technology (or some adaptation thereof).[2]

It is of course less likely that the non-core sector labour force will have housing built by the managers of industry. The industrial capitalist could regard such labour as contributing only to a pool or reserve from which he may be able to recruit, if business flourishes on the upturn of the business cycle, without incurring inflationary shortages. He could also, however, see the necessity of having a local labour force that provides the goods and services at low prices for his core-sector workers. There are indeed many reasons why industrialists benefit from the social stability and economic equilibrium in the local town or city. For this reason they are often concerned to promote welfare in the form of social or sporting events for example.[3] But building well-serviced housing is prohibitively expensive and has to be linked more directly to productivity enhancement in the core labour force. In the steel towns only 40 per cent of the initial workforce was given pukka housing in the official townships (Prakash, 1969), though some of the remainder were provided with reasonable facilities in the neighbouring bustees where they had to build their own dwellings. The manufacturers of food and goods and the tenderers of services for the core-sector labour force, and for the needs of each other, would be the predominant, but not exclusive, occupiers of substandard accommodation usually denoted as slums.

The story we have told so far may seem unrealistically organized and rational. As we have noted in our essay on migration there was a time when migration to cities was regarded as a 'blind inexorable flood': there may still be those who believe the same. A more cautious thesis was developed by the economist Michael Todaro (working on Africa) who argued that since two migrants always come in hope of every one new industrial or government sector job, there would always be a 'surplus' population in the city, shining shoes or selling street foods, while lying in wait at the gate of the factory for a lucky day (Todaro, 1969). In actual fact the Indian evidence suggests that there is very little upward mobility of this kind (Deshpande, 1983); and there is no evidence that the employment problem is any worse in the towns than it is in the villages. Indeed with increased capitalist investment in agriculture, evidence from the north-west of India shows that seasonality of labour demand becomes intensified resulting eventually in capital substitution by capitalist farmers (Byres, 1981). At about the same time the view was widely held that a proportion of slums were temporary accommodation created by the newly arrived migrants as a 'bridge-head' for entry into the mainstream of the city's economy; the architect John Turner recognised this function in the inner-city shanty settlements of Latin American cities (Turner, 1968). Again the Indian experience suggests that such cases are rare, and a careful study by Meera Bapat in Poona covering both inner and outer city slum settlements rejected any notion that hutments were seen as temporary by their resident communities (Bapat, 1981). More realistically they themselves knew that this was all they could afford or the city provide, and their aim was to upgrade their current accommodation as they accumulated the means to do so in the occupations they expected to have to keep for the rest of their lives. Bapat's study, by following households through time, eloquently indicated that neither was there much upward mobility of the Todaro type (and what there was was matched almost equally by downward mobility), nor was there physical movement from poorer to better housing stock, from shanty settlements to official housing colonies for instance. Furthermore we have shown in another essay here, by following through census data over the decades, how, with very few exceptions, migrants prefer to form family units in their urban residence when the physical opportunity allows (Chapter 4); nowadays 'split' nuclear families are largely temporary phenomena. With all this evidence before us in the 1990s we should begin to think of the urban populations as relatively productive and settled. Indeed there are improvements to be made towards both equity and efficiency in the functioning of labour and housing markets, and in the activities of industry and the state; but drastic solutions such as ignoring or removing portions of the urban population are not even vaguely sensible.

In writing of housing as a means to reproduce the labour force, we have suggested that the sector engaged in manufacturing or trading local wage goods or in providing local services has a low priority. But what the industrialist will not do, the state may step in and do, subsidizing the collective activity, as it were, of the industrial enterprises (private or public). Working class housing in general has claims on the attention of the state for other reasons also. It may be regarded as part of the environment of the middle class, which may include the industrialist but only in his role as a personal consumer of urban services rather than as the manager of a labour force. The urban middle class retains the tendency described above to regard urban villages or slums as an eyesore (while rural villages or slums would be romanticized). More sensibly they are offended by the results of poor sanitation which is often just across the street. This is a 'negative externality', as the economist would say, which they are often willing to mitigate at some expense to themselves, that is by bearing the necessary taxation to fund the municipality's capital investment programmes that will improve the housing conditions of the poor. The alternative, which is sometimes too readily thought up, that is physically to remove the offending slum, has the disadvantage of removing the labour force on which directly or indirectly they need to rely. As we have indicated in another essay (chapter 7), this raises the price they have to pay for labour services and for locally traded goods.[4]

Furthermore, in some cities at least, the urban working class has become politically more articulate in the last ten years or so: with their votes being wooed by populist political parties (which are ready to make concessions to labour while continuing to represent the underlying interests of capital) they have at times been able to swing municipal policies towards slum upgrading (and not to their own financial cost).

So far we have looked at the factors governing the economic demand for lower-income housing; these have comprised the purchasing power of the labour forces themselves, the readiness of the industrial employers or their allies in the state to subsidize such housing in order to enhance productivity, and the willingness of the middle class to pay for a more wholesome environment. We now turn to the supply side of the housing question.

Housing Supply

On the supply side we have to remember that the production of urban housing has become an industry like any other. So long as urban incomes retain their current inequality, the urban middle class and elites form a large enough sector of demand for commercial housing to absorb the productive capacity of those enterprises in the construction industry. It has indeed been argued that in the 1980s the richest ten per cent have come to provide a sufficiently large market for consumer durables in India (which would include housing) to falsify any prognostication of declining profits due to lack of a broad-based demand. The purchasing power of this segment of the population in the large cities at least is far greater than surveys may reveal; for it is believed that black money finds its realization in the housing market. Hence the construction industry has no incentive to incur the substantial research and development costs that novel forms of housing and infrastructure for the urban villages of the poor might require. In all, formal sector housing gets supplied by the market for the elites, the middle class, and a tiny fraction of the working class only; in the metropolitan cities this amounts to about 60 per cent of the population, and the proportion diminishes as one proceeds down to small towns and villages (where the market, in the sense of the organized construction industry, provides virtually nothing). In fact the process of concentrating on the rich seems to have intensified in the 1980s: the estimated proportion of the population living in the slums of India's major cities seems to have increased over the previous decade, and the censuses have indicated increasing occupancy per dwelling.[5]

In a land-scarce economy new investment in housing has a positive land cost. In terms of space alone, the social requirements of a steel town for example amount to the building of 100-odd villages of 2,000 population size each (within the first fifteen years of the town's existence). To minimise land costs, population densities higher than those in the villages (which were initially conceived when land was less scarce) are desirable. The equilibrium population density of course depends on the quality of the land, but it should be noted that true land values (i.e. costed in social rather than market terms) may rise in a new-town if the agriculture on the periphery is intensified and channelled into the production of high value cash crops (like vegetables) for consumption within the town (Wadhva, 1983): this will raise the equilibrium population density in the town. Higher population densities, however, or even the equivalent of rural village population densities, require vastly different social infrastructure if a constant level of welfare is to be maintained over time. Briefly put, this is because the environment is incapable of disposing of the domestic waste or providing water for a large localized population (the threshold size depending partly on the local ecology) at these densities without the installation of latrines, drains, garbage disposal, and piped water facilities. Small villages can in theory cope.[6] These land development costs usually double the cost of providing a basic dwelling, even if the latter costs are kept low by the adoption of a rural technology: indeed sites and services alone worked out at Rs. 5,000 to 6,000 per unit in the late 1980s before any dwelling was constructed above foundations.[7]

In practice the urban land market has sometimes come to be regarded as the major obstacle to improving the housing of the poor (Bapat, 1990). Like the property market itself it has often been the forum for shady deals involving huge sums of black money. Whether either of these markets is subject to restrictive practices as such needs investigation in particular cities. Land prices have often risen far faster than the overall price index—at double the rate in Bombay during the 1980s (Pugh, 1989). It is not easy to tell whether this represents under-valuation at the start of the period, an equivalent rise in the productivity of the city's industry and commercial services, or an escalation of monopoly rents (on land or commerce or both). But malpractice in the form of illegal land use and the manipulation of records on floor space are well evidenced. There is no doubt also that huge differentials in price between central and peripheral localities in the city have arisen. In Dur-gapur the price of land in the central commercial area (Benachity) was reported in the 1980s to have reached that in Calcutta; yet the outskirts of the town and indeed some areas within it are used for rough grazing. General amenity value adds a cross-current to differentials of this sort: in Bombay the western seaboard with its refreshing breeze commands a far higher value than the eastern localities which used to swelter in the smoke-laden humidity in the lee of the textile mills, and are still envi-ronmentally inferior. The result is a fairly clear differentiation by class and by housing quality also. Similar differentials characterize Calcutta, to take a third example, with a *tenfold* difference between the price of middle class housing in parts of the north and parts of the south; this is despite very little apparent difference in amenity value but rather more in the prestige attached to living in the south rather than in the north. Intelligent price discrimination on the part of property owners may milk away the consumer surplus of the (fellow) rich in cases such as these; they are not in themselves evidence of malpractice in the sense of market fixing. Where there are huge differentials in incomes (earned and unearned) there will be huge differentials in purchasing power, and hence in market-clearing prices for assets of differentiated quality: small differences in the latter will command large differences in price.

Towards the Equilibrium

This discussion brings us to the conflict of interest that has occasioned the urban housing problem. For we have shown how the industrialist is prepared to subsidize his core-workforce's housing up to a point to maintain good working relations and sustained productivity; and that middle class residents are prepared to contribute to housing for the poor, via taxation, up to a point also, to enhance the local environment. At the same time landowners and private property developers and construc-tion firms have not been prepared to supply land and housing sufficient to allow that market to clear. The resolution of the problem has been attempted as follows. To some extent it seems the state has been involved in bridging the gap. The rationale for so doing is no doubt hazy and developing over time, as has been the role of the state in general. It would be inherently rational economic policy for the state to organize the socializing of some of the common costs of capitalists in the city, and of the middle class residents, taxing both parties to pay for its lower class housing construction. If it went further and engaged in the confiscation of private land, it would be engaging directly in a power conflict. Not surprisingly therefore this has been one of its least practised activities. There have been political cross-currents too. State governments (of one political colour) have cut across the activities of municipal corporations (of another political colour). There is evidence from Bombay, for example, of the State revising its land-use designation in order to compete with the Municipal Corporation for electoral support from slum-dwelling populations. Housing and its attendant amenities have become the tools of patronage and populism, as has been well documented in Madras (Blomqvist, 1988). Our focus here, however, will be on the economics of the problem.

The public authorities have often purchased land, earmarked it for further development, and then left it vacant: hence no private developer or co-operative can acquire that land for low income housing. The inevitable squatting that takes place on such land eventually becomes legitimized, with the resulting complications of replanning that become necessary when drainage and water are supplied, all of which could have been organized or prepared for in the first place. Furthermore, Low Income Group construction loans available from public funds were only permitted to finance up to 75 per cent of the construction costs, the total of which had to amount to at least Rs. 15,000 per unit to be eligible (in 1970s prices; say 35,000 by the late 1980s). What was clearly required was for rather lower cost schemes to be implemented on public land. The Housing and Urban Development Corporation had expended Rs. 2590 million by the mid-1970s, but only Rs. 40 million (1.5 per cent) had been for sites and services schemes. As time went on, it was not as though funds were in theory lacking, least of all in the thriving new-towns built on the industrial boom in western India; but there was often no appropriate authority to mobilize them, or even when such an authority eventually got established the funds were kept aside, like land, for more spectacular purposes than co-operative housing. The new-town of Pimpri-Chinchwad in Maharashtra is an outstanding example (Bawa, 1984), flush with funds reserved inter alia for the building of a stadium. In the case of these rapid growth towns the effect of keeping funds idle is to encourage unplanned development of a socially costly nature: again the case of Pimpri-Chinchwad comes to mind, where unauthorized housing severely encroached on the main Bombay-Poona highway.

Official attention in the 1970s began to dwell upon (or at least hover over) some of the contradictions mentioned above; at least that was the impression the government wished to give. In 1972 the Working Group on Slums reported to the Planning Commission on its fifteen years of slum-improvement schemes (GOI, Planning Commission, 1972). It is indicative of the failure (intentional or otherwise) to grasp the dynamics of the problem that the Group's terms of reference were confined to cities of a population of 300,000 and above. To the credit of the Group, they did point out this anomaly: 'slum growth is not exclusive to Metropolitan Cities and is prevalent to a varying degree in almost every town.' The point can be illustrated well by the fact that by the time of the 1971 census the informal housing stock in the new steel towns had reached 40 per cent of the total housing, but in 1961 none of these cities had reached anything approaching 300,000 in population size, so that concern with slum growth would, according to the criteria above, have been considered premature. Similarly, the deterioration of housing stock of a once more substantial quality and the continuing low level of provision of basic amenities afflict cities that have now almost ceased to grow but are currently below the 300,000 threshold; their population's quality of life continues to lag behind the rest along with their incomes and taxable base. Provision of basic facilities also tends to be regressive as we have further indicated in an essay on the management of cities in this volume (Chapter 7). In Table 8.2. we have assembled data provided by the municipal authorities to the census authorities in 1971 on local expenditure and amenity provision, 'protected water supply' being taken to illustrate the latter. There are problems with the coverage and accuracy of these data, but some general inferences can be safely drawn. The standards of amenity provision deteriorate inversely with town size; yet the smallest town class sizes contain towns that are far larger than the threshold village populations that might be able to deal with fresh water supplies on an ad hoc basis without having to raise community-level resources for social expenditure.

In 1975 a further committee reported to the Ministry of Works and Housing on slum clearance and environmental improvement schemes (GOI, Ministry of Works, 1975). The latter programme has been transferred to the States' administration, which is an advantage given the lack of municipal authorities in several of the new-towns and the lack of an adequate local tax base in many of the towns and cities that have gone into economic decline. The recommended policy, which would appear to be approaching the possible, is fully costed out as follows: land and service provision at Rs. 1,850 per unit in 1975 prices, and materials provided for do-it-yourself construction worth Rs. 150, totalling Rs. 2,000. In practice by the late 1980s a serviced plot of 25.6 square metres costs around Rs. 6,000 (Bapat, 1988). For the first time there is a recommendation to prepare and use hilly and low-lying land, and even land earmarked for parks. As our calculations above suggest, these costs are still a little too high for a complete housing of the incremental urban population.

So far the discussion has focused on the problem of housing the new additions to the urban labour force and population. The problem of the ill-housed backlog still remains, however. The Slum Upgradation Programme dates from 1985-86, and is an advance over the Slum Improvement Programme that dates from the early 1970s in that it allows reconstruction in situ, and permits loans for dwelling improvement, starting at Rs. 5,000. Despite the improved sanitary standards which it lays down, the infrastructural expenditure averages only Rs. 2,000 per household. It is not a *sufficient* solution to the backlog problem, however. For example, of the 331 slum pockets in Bombay that were situated on land owned by the State government (or acquired by the State's Housing and Area Development Authority), only one half were found suitable for upgrading. Of the rest, one half were in such precarious locations that implementation of the programme was deemed technically infeasible (or at least extremely costly), and the rest were ruled out as incompatible with the Development Plan (Sinha, 1989).

None of these supposedly comprehensive programmes address the question of the deteriorating standards in the older towns or the declining towns. Only locally has the problem of dilapidated tenement blocks been reviewed (for example, in Bombay: Kerkar et al., 1981). And the question of the pavement dwellers in the metropolitan cities remains unanswered. The assumption is that they will need relocating, and will not be part of an upgrading scheme. But the numbers of households involved are huge, and their livelihoods are intricately tied up in the local economy, making relocation economically problematic (and in some places politically impossible).

Finally, the assumption is still made that somehow the land required will be acquired at a moderate cost. It is well known by now that in practice this has not been the case. The Urban Land Ceiling Act has failed to bring more land on to the market or into the hands of the state; if anything it has brought less. Municipal authorities have indeed favoured marginal land for rehousing schemes, but have often failed to make such land habitable: in the mid-1970s Bombay instituted a large slum clearance scheme, and relocated people on what were supposed to be prepared sites (these included distant, marshy, and polluted areas like Deonar), some of which had not been prepared at all. Municipal authorities have also been ready to move people from sites where they may be an eyesore with a possible effect on property values in the vicinity (Bapat, 1988).

The Possible Equilibrium

In our concluding scenario we too will make the assumption that the state will acquire the necessary land for its housing schemes. In fact much of the necessary land is already in the hands of the central or state government departments. This is particularly often the case in industrial new-towns. Much again, however, is privately owned, and liable to be 'invaded' if left vacant and obviously surplus to current productive needs. The National Housing Policy of 1988 envisaged taking steps to prevent the speculative holding of vacant plots and to pre-empt price rises in areas of urban development. But there are powerful forces that will undoubtedly seek to forestall the wholesale appropriation of their economic rent in this way. Nevertheless it would probably be in the interests of all parties if some of that land were released for the organized planning of serviced residential communities. For if cities are to develop rapidly, then labour and congestion costs have to be kept down; restrictive housing strategies ultimately make labour scarce, and strangle the very economic growth on which much of the capitalist's profit and rentier's rent depend. Only a small minority can make a mint out of a shortage economy, because it is a zero-sum game. It is part of the logic of the World Bank's interest in assisting the housing programmes in metropolitan and some of the new industrial cities that this is recognized.

In general terms our argument for an equilibrium solution runs as follows. In cities growing at about four per cent per annum (which was the average rate of urban growth in the 1970s) roughly two per cent is added annually consisting of people who are unable to afford the economic return (interest or rent) even on housing which is designed for the so-called 'economically weaker section'. The capital cost of such housing amounts to at least Rs. 15,000 in late 1980s prices including land and services on the outskirts of a large city; this implies an annual rent of Rs. 1,200 to give an acceptable annual return on public capital (eight per cent say). Even if as little as ten per cent of a household's income is allocated to such a purpose, the monthly rental charge (or initial interest) implied by these figures means that only those households in the top 60 per cent of the urban household income distribution in the more prosperous cities could afford to pay.[8] In the poorer cities probably only the top 40 per cent could qualify. Our conclusions are that, unless there are to be rent subsidies, or subsidized interest charges for home owners, up to one half of the incremental population (that is two per cent of the four per cent in urban growth) has to be accommodated on open developed plots costing initially about Rs. 6,000 (excluding building material costs). That in itself is a startling indicator of how costly a city is.[9]

But suppose rent or interest subsidies are allowed and that the subsidy can somehow be paid so that the problem of affordability outlined above is mitigated. The question still remains as to whence the capital is to be obtained to finance construction costs of Rs. 15,000 per unit for housing the incremental lower income population. The expenditure required in a city of currently 100,000 population adds up to: Rs. 15,000 times 2,000 (the incremental population in the first year), divided by an average household size of five, which equals six million rupees (or nearly Rs. 60 per head). This would exhaust between one third and one half of a municipal budget in a class I city (depending on its prosperity, as examples from Maharashtra (excluding Bombay) would seem to indicate (after allowing for a fourfold increase in prices since 1971); see for example Table 7.3. In India housing finance is obtained from agencies other than the municipalities, especially those administered by the state governments like the Housing and Urban Development Corporation. But this simply means that the finance has to be transferred from that raised elsewhere. The drain on investment funds can be seen when we point out that the requirement sketched above amounts to between ten and fifteen per cent of the national savings. This is an unprecedented allocation to the single sector of urban low-income public sector housing, and only meeting half the requirement at that (when the public housing sector normally absorbs less than five per cent of planned investment outlay, and the whole of the social services sector was intended to absorb no more than 20 per cent of the central and state governments' expenditure in the Seventh Five Year Plan). Something like one-half of that figure (e.g. the Rs. 6,000 implied by sites and services) would clearly be more realistic.[10]

In all this discussion we have deliberately omitted from our financial estimate the complicating addition of urban re-construction (upgrading of slums and dilapidated housing); and we have only concentrated on additions to housing for the poorer half of the incremental population. The ultimate outcome of all this is to prove that, wherever the finance is to come from, it is prohibitively costly to allow for more than open-developed plots to accommodate half of the four per cent annual urban population growth.[11]

These striking conclusions should be hardly surprising in a sense. For most of the migrants were accustomed to build their own houses on plots that were unserviced in the villages from which they came. It is simply a realistic reflection of the fact that India is a relatively poor country, and a city like Bombay or Madras cannot, for all their modernity, pretend otherwise. Owing to income inequalities the rich may be able to afford apartments comparable to those in New York or London, but the poor cannot afford the equivalent of London's council housing. Nor indeed can the Municipal Corporation of Bombay, the State of Maharashtra, or the Government of India together afford to build to such standards, except to accommodate a small proportion of the core-sector working class in the more productive industries. The city of Madras tried a more comprehensive programme of this kind for a while, but it was quickly brought to a halt.

Why should there be so much prejudice against the open developed plots, or their equivalent, the upgraded shanty settlements or bustees? With proper planning, including that of community organization to ensure adequate maintenance of facilities and collection of appropriate charges, such settlements are both hygienic and aesthetic. They simply come to resemble well-kept and well-serviced rural villages. They can be adorned with trees, bushes, and potted plants. If the reader regards this as a new romanticism let him or her wander observantly through some of the outer suburbs of Bombay (in Vikhroli for instance) or some of the upgraded localities of Calcutta (Cossipore, for instance) or the fringes of Benachity in Durgapur. There are urban pockets in all cities where this has been achieved. What one has to envisage, and it does amount to a vision, is a vast urban village of this kind. It is the only economic solution to the problem of accommodation that enables towns and cities to continue to grow in equilibrium. The alternatives, whether insanitary slums alongside piecemeal rehousing, or banished populations and stagnating urbanization, are unnecessary, undesirable (except to a minority), and unjust. Ultimately they tend to be to the disadvantage of rich and poor alike.

NOTES

[1] Rs. 1,000 was the compensation offered in the late 1950s by Hindustan Steel to the villagers whose homes were destroyed to make way for the Rourkela plant (about Rs. 15,000 in late 1980s prices) (GOI, Registrar General, 1961). But any valuation of rural assets is extremely hazardous. It would be better to argue on the basis of the cost of constructing a rural-style dwelling in an urban area: the simplest pukka building costs about Rs. 8,000 to 9,000 subtracting land development costs. The two estimates are consistent.

[2] The author and M.Bapat carried out a survey of household incomes and occupational remuneration among Pune shanty dwellers in 1988 (Bapat, Crook and Malaker, 1989). The figures we produce here are at least consistent with the findings from that survey. For example, the poorest ten per cent of households had a yearly income of approximately Rs. 3,000. If they could afford ten per cent of that for rental or interest plus repayment charges, they could afford Rs. 300 towards housing. Repayment plus interest on a HUDCO loan for serviced plots came to about eight per cent. This implies that these households could only afford to incur a capital expenditure of Rs. 3,750. A similar calculation suggests that from the poorest 25 per cent to the richest 25 per cent the range of affordability goes from about Rs. 8,000 to Rs. 15,000 (assuming here that 15 per cent of household income could go towards the necessary charges).

[3] The following is a candid comment by the Nayar Committee on Jamshedpur: 'A choice must be made as between a growing expenditure on town services or labour discontent on vital matters,' quoted by K.C. Sivaramakrishnan (1976), who also says himself that welfare services are provided to ensure a healthy and contented workforce on assumption that this follows. Being a former Secretary of the Durgapur Development Authority he should know.

[4] Meera Bapat's survey (referred to above) also showed conclusively that dwelling in proximity to place of work was essential for the urban poor (Bapat, 1981): transport in the large cities was prohibitively costly in time, effort or money; (if heavily subsidized then it becomes grossly overcrowded, reducing the energies and hence productivity of its customers).

[5] In Bombay the percentage is thought to have gone up from 35 per cent (in 1976 when a survey was carried out) to over 40 per cent in the 1980s (the figure now quoted in planning documents, though on what basis is not clear).

[6] This is not to argue that villages are inherently healthy places. With settled agriculture, disposal of waste becomes a matter of concern if diseases like dysentery and hookworm are to be avoided. But knowledge and organization are alone sufficient to maintain levels of welfare in this case. A further domestic pollutant comes from the burning of soft fuels: again costs are incurred if this problem is to be mitigated in cities, whereas in villages it is unlikely to be much of a problem outside of individual dwellings.

[7] These data are based on experience in Poona in the late 1980s. In

comparison it has been estimated that relocation of a household in Bombay would cost Rs.7,500 for a developed site of 27 square metres, plus another Rs.7,500 for a temporary structure on top (Sinha, 1989).

[8] The rent implied in the figures above is Rs. 100 per month; household income would have to amount to Rs. 12,000 per annum to bear this. The official urban poverty line was set at Rs. 7,300 in the mid-1980s. This suggests that all of those below the poverty line would be unable to afford housing designed for the economically weaker sections. There are various estimates of the proportion of the urban population below this and other poverty lines: approximately 35% would seem to be realistic (Harriss, 1989). Whether we assume that the accommodation will be owned or rented makes little difference to the essence of the argument here. Owners will face interest and repayment charges that are higher to begin with but diminish over time. The incentive to sublet is strong and increasingly apparent. This puts the intended beneficiaries back in the status of squatters or slum dwellers. However, a fully comprehensive scheme, by putting enough housing on the market for everyone to afford, would eliminate the slum problem, even if subletting did exist. Clearly by positing ten per cent as the maximum acceptable proportion of household income to devote to housing we are making a value judgement. When we are talking about households that spend between 70 per cent and 80 per cent of their incomes on food, it seems reasonable to suppose that an extra expenditure of ten per cent could perhaps be accommodated without harm, but an addition of 20 per cent would almost certainly cut into food expenditure with undesirable nutritional consequences.

[9] This scenario is kept simplistic in order to sharpen the basic point. Other essays in this book have been at pains to point out that much diversity exists, and that the sustained four per cent urban growth really only describes the experience of a limited number of cities. If a town is growing very fast, it is true that initially housing may be shared by in-migrating males, thus reducing the immediate infrastructural requirement. But subsequent family formation is both socially desirable and often encouraged by employers. In any case, subsequent migrations tend to be of complete nuclear families. Nowadays in India it is only a very small proportion of nuclear families that remain permanently split. It would therefore be sensible for urban authorities to plan to accommodate a household of five or so for each independent migrant.

[10] By contrast the private sector is responsible for a considerable financial investment in housing: added together, public and private expenditure on permanent residential construction amount in total to between fifteen and 20 per cent of gross fixed capital formation in India (UN, 1989). In urban areas most privately financed residential construction (sufficiently permanent and legal to get into the records) would be for the upper middle class and elites, the rest being private company housing for core-sector labour (as at Jamshedpur). The figures therefore tell us something of the skewed nature of income distribution.

[11] In this light the proposed rehabilitiation scheme for Bombay slum-dwellers outlined in the Kerkar report (1981) is idealistic and not replicable

elsewhere. Of the 800,000 families requiring rehabilitation (about ten per cent of the population when the report was written), it is assumed that only fifteen per cent would opt for sites and services, though, it says, 'in the short run sites and services would be more popular.' When the report laudably carries out a budgeting exercise for rehabilitation on these lines it envisages that only Rs. 2,200 crores of the required 4,200 would be raised by the State of Maharashtra. The balance would come from central government or the World Bank.

Table 8.1
Housing of Factory Workers in Selected Industrial Cities, 1958-9

Characteristics of Accommodation	Percentage of Industrial Workers Reporting Characteristics in Selected Cities						
	I	II	III	IV	V	VI	VII
Chawls/Bustees	35.0	38.7	63.3	46.2	98.3	66.7	59.2
Flats	17.8	42.9	13.3	0.4	0.8	1.1	0.8
Independent buildings	42.8	12.1	7.5	7.5	0.3	10.0	28.3
Other	4.4	6.3	15.8	45.8	0.6	22.2	11.7
	100	100	100	100	100	100	100
Employer-owned	38.9	56.2	88.3	73.7	6.7	3.9	5.8
Self-owned	21.7	22.9	10.8	18.3	72.5	21.7	14.2
Privately owned	38.3	19.2	0.8	6.7	20.8	74.4	79.2
Other	1.1	1.7	0.1	1.3	0.0	0.0	0.8
	100	100	100	100	100	100	100
Permanent kaccha	28.3	35.6	19.2	21.2	44.2	19.4	33.3
Permanent pukka	41.7	57.9	46.7	36.7	33.3	60.6	46.7
Temporary kaccha	16.7	0.4	15.0	6.2	16.7	8.9	15.8
Temporary pukka	12.8	7.1	19.2	35.8	5.8	11.1	4.2
	100	100	100	100	100	100	100
Without latrine	46.1	18.3	90.8	96.7	72.5	23.3	37.5
With private latrine	27.8	33.3	0.0	1.7	10.8	0.0	7.5
With communal latrine	26.1	48.3	9.2	1.7	16.7	76.7	55.0
	100	100	100	100	100	100	100

Note: Principal industrial characteristic of the workforce is indicated in brackets after the name of the town (see Key below).
Source: Government of India, Ministry of Labour, Labour Bureau, *Family Living Surveys among Industrial Workers, 1958-9*.
Key: I Asansol (heavy industry); II Jamshedpur (steel): III Raniganj (coal); IV Jharia (coal); V Monghyr (mixed industry); VI Sholapur (textiles); VII Madurai (textiles)

Table 8.2
Municipal Revenues, Taxes, and Water Supplies According to Size
Class of Towns, 1971

Size Class of Towns	Per Capita Receipts (Rupees per annum)	Tax as percentage of Receipts	Percentage of Towns with Protected Water
BIHAR			
I, II	12.6	40.1	100.0
III and below	9.4	30.9	93.4
MADHYA PRADESH			
I, II	27.6	66.0	100.0
III and below	15.6	60.7	41.4
MAHARASHTRA			
I, II	47.8	62.5	95.2
III and below	13.8	48.4	1.0
TAMIL NADU			
I, II	33.0	48.3	89.7
III and below	15.4	44.4	62.2

Note: Figures refer to years immediately previous to the 1971 census; simple
averages have been taken for groups of size classes.
Source: Calculated from Government of India, *Census of India, 1971*, Bihar,
Madhya Pradesh, Maharashtra, Tamil Nadu, Town Directories.

Chapter 9

AFTERWORD

Policy Implications?

At the end of a book such as this, some readers may ask, 'But what are the policy implications?' I have long come to the conclusion that this question is insufficient by itself: it invites the following and necessary additional question, 'The policy implications for whom?' The answer that often seems to be expected is: 'For the Government.' If one feels this is an important and useful answer, then one has already taken up a position on the nature of the state. It is not, however, the same position that I happen to hold.

I do not have an elaborately developed theory on the nature of the Indian state, nor would this be an appropriate place to discuss it if I had. I do, however, believe that no government has such an autonomy from the social forces that construct the society as to make the answer given above at all meaningful: in the words of the economist, it is pointless suggesting Pareto-optimal conclusions to a government whose interest in Pareto-optimality is minimal.

In Indian industry and the urban sector it would seem that capitalism as the major mode of production is fairly well developed, and that the existence of class division and consciousness are fairly well testified. This is not to deny the ambiguity of some of the state industrial enterprises; nor to overlook the evidence that capitalist industry has been taking strides to put back the clock, to divide and deprofessionalize the workforce, to minimize the size and impact of a self-conscious class of wage labour (as the steady growth of subcontracting bears witness). The position in the urban sector is complicated further by the landowner, who may or may not be the same as the government or the industrialist; and by well-defined and powerful fractions of capital such as property developers. Beyond which much debate exists as to whether one can describe the rent-seekers, who milk consumer and producer surplus in return for legal and illegal services, as a self-conscious 'class' in the Indian society; that they are a large and active component in the economy and society is not in doubt. So far we have argued there are fundamental forces of class penetrating urban society; furthermore, these are partially cemented by the age-old cohesion and linkage of caste. If these are fundamentals, there is additionally a plurality of social interest groups, whose attachment to the fundamental social forces can no doubt be traced, but whose day to day activity may be circumscribed by more parochial objectives. I refer to various quasi-independent bureaucracies, to political groups and parties, to non-governmental organizations of a charitable or activist nature. They are the waves on the surface beneath which the fundamental currents move.

The implications that derive from the analysis in this book would be different for each and every one of these classes, groups, and actors. In which case it would be less misleading to refer to the implications as 'tactical' or 'strategic' rather than 'policy': 'policy' seems to suggest government. Let us give some examples. The industrial capitalist might feel I have made a case for the role of urbanization and industrialization in forwarding the course of economic development, and for the strengthening of urban infrastructure and public price reforms toward that end; he might use this argument to counter urban 'middle class' (or rural 'middle peasant') views that de-centralization is desirable per se, and to win vital urban votes from the former. The trade unionist might feel I have indicated the risks that lie in a faltering in the momentum towards the formation of a mature working class, as a result of the steady substitution of capital for labour, and more importantly in the increased recourse to contract labour. More specifically, in Chapter 5 the message may be seen to be one of working towards more national or regional links between labour organization and income support; to prevent the emergence of economic islands, especially when these are single-industry towns, which favour the perpetuation of weak labour and strong capital. The low-caste working class activists might find facts and figures herein to strengthen the argument that, despite the political manoeuvring around the reservation of jobs in government service for the Scheduled Castes and Scheduled Tribes, the penetration of these people into the organized, better-paid, and more politically conscious industrial wage labour has in many important places been minimal. The discussion herein on the virtues of a planned urban environment, where the improvement of living conditions for the poorer half of the population by sites and services and slum upgrading schemes turns out to be the only practical solution to the problem of urban environmental degradation, could appeal both to the middle-class voters and certain of the state bureaucracies responsible for the urban sector. One has to remember that governments or state authorities at different levels in the administrative system do not always each have the same objectives in mind. Unlikely allies such as industrial chambers of commerce seeking to attract business to their city (as they once had to attract labour), and activist community leaders in the slum colonies (seeking to satisfy their fellow members) may find a common interest in the economic logic of our argument on the housing problem. In a not dissimilar vein runs the argument that wage labour may still expect to win some concessions from capital towards improvements in health in high productivity industry, but that generally the working classes must increasingly expect to gain environmental improvements from pressurizing the representatives of the urban middle classes who see their own interests lying in the same direction. The special financial (and political) needs of the 'new-towns' (some of which, like New Bombay, are still being created, while others are now forty years old), and of the stagnating towns, that we have indicated, might appeal to various local lobbies, including Development Authorities and Municipal Corporations, where they exist. The implications, whether from facts, figures, or analysis, in this study will be different for different social actors, of which the government, if it acts at all, will only be one.

It cannot be expected, however, that representatives from all these classes or social groups will get to read this book. Its largest readership will inevitably be that of students and academics. One has to have some faith therefore in a process of percolation: that to some extent its content will eventually reach a wider audience. One has also to have faith in the value of knowledge itself. It is ofcourse true that various facts made known in this study can be used ultimately to the common good or to the common bad; to the advantage of the already advantaged, or the disadvantage of the already disadvantaged. I have heard it said, for example, by ardent free-marketeers, that my indication of the demographic-economic disequilibrium that occurs with the establishment of huge steel towns should be an argument in favour of those who would wish to remove the protective tariff on steel, and close down the large public-sector plants; whereas the argument I wished to make implied greater state involvement in subsidizing diversified investments in these localities, an argument more likely to appeal to the socialist. My view on this is that, although knowledge can be turned to use for good or evil, ignorance is far the worse potential evil of the two. In this sense knowledge is like a competitive industry. The more there is, the more one needs, to avoid being outwitted by those who have the most. The only regulation required is that knowledge should be got from well-researched materials and should be carefully presented, that is it should not be false or seriously biased: though in the last resort the selectivity of the presentation, and the personal stance of the communicator, cannot be avoided.

This then is my answer to those who would ask, 'What are the policy implications of your work?' Read, and see if there is anything that helps you in what you see as your role in society.

A Summary?

At the end of a set of essays, some readers will demand a summary; but essays are self-contained, and a summary not altogether appropriate. We will none the less attempt one.

India's urban demography has been characterized by diversity over time and space. For thirty years or so from the late nineteenth century (when the censuses began) to the early twentieth, some cities prospered, while others went into decline (Chapter 2). The high mortality of the period kept natural increase close to zero, so that the major contributor to city growth was migration. Social forces no doubt kept labour underemployed, both on the land and in the new industrial centres; but overall the labour supply was unable to grow until mortality began to fall in the 1920s (and then most of the increase was delayed until the surviving children reached adulthood), and more precipitously from the 1940s (Chapter 6). For this reason among others most cities did not readily reach the threshold of economies of scale; and arguably more rapid industrialization could not have taken place.

While demographic constraints on the rapid development of industry are arguable, educational and skill constraints there certainly were (Chapter 3). Furthermore, industrialists were wary of whom to employ in increasingly centralized workforces. The massive step-up in labour demand for the industrial development of the 1950s meant that labour recruitment took place on an unprecedented scale. It has not been adequately appreciated that this labour *demand* was the engine of urban growth, and contributed to the demographic diversity of cities of varying industrial character; further, those migrants sowed the seeds of future natural increase. The notion that cities grew from an unchannelled flood of population movement is false. From the 1950s through to the present time, urban growth has been as much due to natural increase, death rates having fallen well below birth rates, as to migration (Chapter 6). This scenario is illustrated well by the experience of the steel towns (Chapter 4). Huge economies of scale demanded large labour migrations; the migrants were not landless, but came from small landholding backgrounds. Initially the workforce came alone, but they soon reconstituted their immediate family in the city. A further wave of migrants was required to provide the wage goods and services needed by the initial labour force; they usually came from poorer landless backgrounds, and severed their link with the land. Employment for the second generation now poses severe problems in those cities.

It is true that there is generally some security in industrial employment in comparison with rural wage labour; but there are exceptions. Towns of diversified industry may provide alternative employment to those thrown out of work, and households with diversified sources of employment may survive layoffs. But single-industry towns less easily develop those capacities (Chapter 5): households therein have to seek strategies to secure their income base in the event of industrial decline or closure. Such an eventuality is not confined to the western world; with industrial restructuring it has already occurred in Indian mining and textile towns, and even in a recently industrialized locality. The core labour force, having ties with the land or educated children, may survive on a broader income base; the rest, of landless origin and able to afford little investment in human capital, may be unable to do so. Intensified social duality of this nature has not always been appreciated as a by-product of the new industrialization. Similarly, the failure of the new urban infrastructure to extend to the mass of the newly formed urban households has gone unnoticed (though not by those who live in the new-towns themselves); child mortality is indeed low for those who live in the planned quarters of the steel towns, but as high as anywhere in the unserviced areas of Bombay or Calcutta for the 50 per cent of the population whose homes are outside the planned quarters (Chapters 4 and 8).

But if, as we claim, cities contribute to economic growth, there must be a strategy for dealing with their attendant malaise (Chapters 7 and 8). Again the diversity is important. Towns that have lost their industrial advantage, often old towns growing slowly, need as much attention as those that tend to dominate social concern or international investment. For they too have their slums. In the rapidly growing localities the solution is more services, not fewer: their growth indicates their prosperity. Urban authorities need to cream off surpluses from capitalists and rentiers (especially the latter as the land market is often the most distorted) for infrastructural construction, but also to raise finance from within slum communities themselves to pay for their use of the new infrastructure. But the sheer economics of the question dictates that *most* of the house-building has to be done by the home-owners themselves. Checking urban growth by restricting migration, industrialization, or spontaneous housing, will eliminate the benefits of cities along with the nuisance. Polluting industries and congesting traders need to be taxed or controlled, not driven away.

Conclusion?

At the end of this analysis of the interrelationships between industrial and demographic processes, as they have occurred in the comparatively recent history of India's industrial cities, one is left with a slight sense of surprise. Some of the contradictions between industrial and demographic requirements have been met with an unexpected resilience. The social and economic fabric of the cities has remained durable. There is a crumbling veneer, to be sure, and even some structural weakness, but no general collapse. This is contrary to the expectation of many observers from outside India, and even from within. It may be contrary to some of the implications of the arguments made even within some of these essays. The fact that the urban structure holds together, the system works, industrialization and urbanization are sustained through the decades since Independence, does not mean, however, that the process has been acceptable to more than a minority, or even that it has been civilized. Great hardship and sorrow on the one hand, and profligate waste and misuse of resources on the other have been the prices paid.

We are making a case here for optimizing under certain political constraints. It may be that agriculture and rural infrastructure have been starved of resources; if that is true, to restore the balance is a macro-political and macroeconomic problem. It is no solution to stifle the growth of the existing cities. But even within the current resource constraints, there is a huge scope for better management of the urban resources that there are.

The immediate effect of some of the planning legislation and public sector pricing policies that we recommend might indeed be to slow the growth of some of the cities. But soon, and as a result, a more equitable, and possibly faster growth of both urbanization and industrialization will follow. And even if at some future date industrial import tariffs are lowered, and marginal resources channelled away from industry, industry and city would contribute more (under our recommendations) per unit of resources utilized, than before, to India's development and growth.

But those who still see the dark side of cities, and would seek to check migration, should ask themselves this question. If the Hooghly runs polluted through Calcutta, do you suppress the source at Gangotri?

Appendix A

THREE CONTRASTING URBAN DEMOGRAPHIC MODELS

The three examples for illustration are as follows:
   (a) Concentrated heavy industrial investment on a new site
   (b) Balanced industrial growth at an existing centre
   (c) Decline of an industrialised locality
For these demographic projection models migration is the control variable, and takes on the values indicated in the first panel of the table A.1 below.

The following parameters are set for all three models:
Initial age structure: all-India urban age structure as recorded in the 1961 Census.
Fertility: total fertility rate of 5.1, with age-specific rates derived from Coale-Trussell models; these are 0.0069, 0.1361, 0.2590, 0.2445, 0.1871, 0.1252, 0.0542, 0.0070 for the eight five-year age groups from 10-14 to 45-49 respectively.
Mortality: female life expectation at birth of 40 years. Survivorship by five-year age groups as in Coale-Demeny model life-tables, West level 9.
Migration: the age structure is constructed from the data provided by the NSS and discussed in Chapter 3.

Further details may be obtained from the author.

TABLE A.1
CONTRASTING URBAN DEMOGRAPHIC MODELS*

PANEL A: Net-inmigration rate for each five-year period expressed
as annual per cent per annum

Model	Quinquennium					
	First	Second	Third	Fourth	Fifth	Sixth
(a)	86.5	1.8	0.2	0.2	0.2	0.0
(b)	1.0	0.9	0.9	0.9	0.8	0.0
(c)	-2.9	-3.2	-3.5	0.0	0.0	0.0

Comment: Model (a) has extreme inmigration in the first five years
only. Model (b) has roughly constant migration over most of the
period. Model (c) has substantial outmigration for the first three five-
year periods.

PANEL B: Difference between the growth of the male labour force
(15-59) and that of the total population for each five year
period expressed as annual per cent per annum

Model	Quinquennium					
	First	Second	Third	Fourth	Fifth	Sixth
(a)	85.4	-1.4	-0.8	-0.8	-1.4	-0.8
(b)	0.5	0.8	-0.2	0.0	0.0	-0.4
(c)	-0.9	-0.6	-1.9	0.1	0.0	0.0

Comment: Positive values indicate that the labour force is growing
faster than dependants (defined as the rest of the population); negative
values indicate that dependants are growing faster than the labour force.
In model (a) the dependants are mainly young, and continue to grow
throughout the period after the initial inmigration (which is predominantly of working-age adults). In model (b) there is little difference in
the growth of labour or dependants. Model (c) has young and old dependants, and their relative growth in the first three periods is partly due to
the outmigration of the working age-groups.

*Note: Parts of the panels are graphed in figure A.1

PANEL C: Growth of labour-force entrants at ages 15-19, for each
five-year period, expressed as annual per cent per annum

Model	Quinquennium					
	First	Second	Third	Fourth	Fifth	Sixth
(a)	186.8	-7.1	-1.6	1.7	-3.3	7.1
(b)	4.9	3.3	-0.9	3.1	1.4	1.6
(c)	-0.4	-0.5	-6.9	6.5	1.1	1.1

Comment: Model (a) shows severe fluctuations in labour entry long
after the initial inmigration. Model (b) has a steadier profile. Model (c)
has substantial outmigration of the labour entry cohort; once this stops
there is considerable growth in the new labour entrants as the young
dependants age.

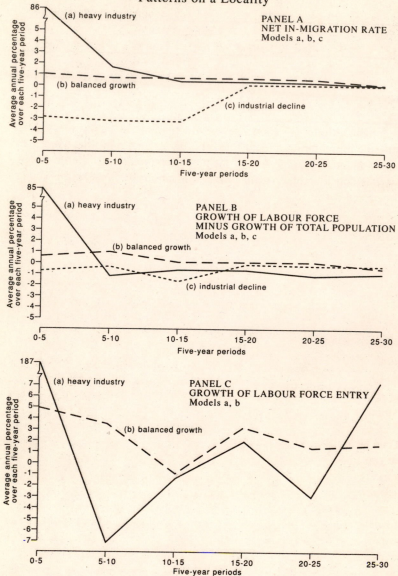

Figure A. 1.
Simulation of Demographic Effects
of Three Contrasting Migration
Patterns on a Locality

PANEL A
NET IN-MIGRATION RATE
Models a, b, c

(a) heavy industry

(b) balanced growth

(c) industrial decline

Average annual percentage over each five-year period

Five-year periods

PANEL B
GROWTH OF LABOUR FORCE
MINUS GROWTH OF TOTAL POPULATION
Models a, b, c

(a) heavy industry

(b) balanced growth

(c) industrial decline

PANEL C
GROWTH OF LABOUR FORCE ENTRY
Models a, b

(a) heavy industry

(b) balanced growth

Notes: Labour force is taken as aged 15-59, the rest of the population being defined as dependent. Labour force entry age group is taken as 15-19.

Source: Table A.1.

Appendix B

INTERPRETING LABOUR FORCE STRUCTURE AND RURAL URBAN DISTRIBUTION OVER TIME

Let us make the following definitions for using the census data:

URBLABAG: The proportion of the urban labour force that is in agricultural work

RURLABAG: The proportion of the rural labour force that is in agricultural work

LABAG: The proportion of the total labour force that is in agricultural work

URBPOP: The proportion of the total workforce that is urban

RURPOP: The proportion of the total workforce that is rural

From the data in the 1971 Census we obtain the following:

URBLABAG	URBPOP	RURLABAG	RURPOP	LABAG
(0.11	x 0.18) +	(0.82	x 0.82) =	0.7

For 1901 we have the following (under the same heads):

| (a | x 0.1) + | (b | x 0.9) = | 0.7 |

where a and b are unknown, and the proportions of the workforce that are rural and urban have been approximated by the proportions of the total population that are rural and urban respectively; this probably underestimates the urban figure slightly.

It can be easily checked that if b > or = 0.82 then LABAG > 0.7. Hence we can state immediately that there must have been fewer agricultural workers (cultivators and landless labourers) as a proportion of the rural labour force in 1901 than in 1971.

However it is not possible to make any conclusive statement about the proportions of the urban labour force in agricultural work. They may have been more or less in the past or have remained the same as the following two simulations will show:

| (0.2 | x 0.1) + | (0.75 | x 0.9) = | 0.7 |
| (0 | x 0.1) + | (0.77 | x 0.9) = | 0.7 |

In fact there is a graph in the 1901 Census Report that suggests that the value of a should be 0.1; hence b comes out at 0.76:

$$(0.1 \quad x \quad 0.1) \quad + \quad (0.76 \quad x \quad 0.9) = \quad 0.7$$

This is the basis of our claim that, in some sense, non-agricultural work has become increasingly urban as the century progressed. A similar exercise could be done for the workforce employed in manufacturing. The estimates indicate that 8 per cent of the rural labour force was in manufacturing in 1901; the figure for 1971 is 5 per cent. For the urban labour force the 1901 estimate (again from the graph) is 19 per cent; the 1971 figure is 28 per cent. So manufacturing became more concentrated in urban areas. The changes are great enough for one to hypothesise that the re-drawing of urban boundaries to encompass previously 'rural' industry cannot have been sufficient to affect the argument: typically this occurs where there are agglomerations of rural industry—which makes our point.

Finally some scholars have detected a change between 1971 and 1981 slightly towards the de-centralisation of non-agricultural activity (Mills and Becker, 1986). We give the figures without comment, the changes being too small for confident analysis:

URBLABAG	URBPOP	RURLABAG	RURPOP	LABAG
(0.11	x	0.21) +	(0.81	x	0.79) =	0.67

Appendix C

REGRESSION MODELS

(i) The Relationship between Male and Female Migration

The following linear regression result was obtained, using 26 un-weighted observations from the urban populations of the following districts in the eastern region of India from the 1961 Census: Calcutta, Howrah, 24-Parganas, Hoogley, Midnapore, Burdwan, Birbhum, Bankura, Patna, Gaya, Singbhum, Ranchi, Dhanbad, Monghyr, Shahabad, Bhagalpur, Raigarh, Durg, Bilaspur, Bastar, Mayurbanj, Sundargarh, Cuttack, Ganjam, Bolangir, Baudh-Khondmals.

Female migrants of less than one year duration in the urban area =
   - 96*
   + 0.32* male migrants of less than one years duration in the urban area
   + 0.17* male migrants of 1-5 years duration in the urban area
   - 0.062 male migrants of 6-10 years duration in the urban area.
$R^2 = 0.32$

Notes: * indicates statistical significance at 5% level, but as the sample was a purposive one, drawn to study in depth certain of the heavy industrial localities with a number of controls, this has no rigorous statistical meaning; it does suggest, however, which of the parameters are the most reliable.

Rather similar results were obtained by taking all the districts of Maharashtra as the observations.

(ii) The Relationship between Urban Sex Ratios and Other Demographic Variables at the Level of the State: First Model

Sixteen states of India were included (the exclusions being mainly the small States in the North-east): Andhra Pradesh, Bihar, Gujarat, Haryana, Himachal Pradesh, Karnataka, Kerala, Madhya Pradesh,

Maharashtra, Orissa, Punjab, Rajasthan, Sikkim, Tamil Nadu, Uttar Pradesh, West Bengal.

The following regression results can be reported:
Urban sex ratio (m:f) =   - 0.003691
                              + 1.1586 total sex ratio (m:f)
                              + 0.0002089 urban decadal percentage
                              increase 1971-1981
                              - 0.001896* percentage of class I urban
                              population in total urban population
$R^2 = 0.76$

Notes: * indicates statistical significance at the 5% level.
Class I urban areas are cities of 100,000 population and above.

(iii) The Relationship between Urban Sex Ratios and Other Demographic Variables at the Level of the State: Second Model

Data are for sixteen states as in first model; the following results can be reported:
Urban sex ratio (m:f) =   - 0.009087
                              + 0.2258* sex ratio of migrants (m:f) of
                              less than one year's duration of stay
                              + 0.8986* total sex ratio (m:f)
                              - 0.0003672 urban decadal percentage
                              increase 1971-1981
                              - 0.001248 percentage of class I urban
                              population in total urban population
$R^2 = 0.85$

Notes: * as for Appendix C (ii).

BIBLIOGRAPHY

Arnold, D. (1987). Touching the Body: Perspectives on the Indian Plague 1896-1900, in Guha, R. (ed.) (1987). *Subaltern Studies V: Writings on South Asian History and Society.* Delhi: Oxford University Press.

Bagchi, A. (forthcoming). In Byres, T.J. (ed.) *The State and Development Planning in India.* Delhi: Oxford University Press.

Ballhatchet, K., and Taylor, D. (1984). *Changing South Asia: Economy and Society.* Hong Kong: Asian Research Service.

Bapat, M. (1981). *Shanty Town and City: the case of Poona.* Oxford: Pergamon.

Bapat, M. (1988). Critical Evaluation or Toeing Official Line? Report on a re-habilitation project, *Economic and Political Weekly*, 23:16.

Bapat, M. (1990). Allocation of Urban Space: Rhetoric and Reality—Evidence from recent jurisprudence, *Economic and Political Weekly, 25:28.*

Bapat, M. and Crook, N. (1988). Duality of Female Employment: evidence from a study in Pune, *Economic and Political Weekly*, 23:31.

Bapat M. and Crook, N. (1989). Behind the Technical Approach to Slum Improvement, *Waterlines*, 8:1

Bapat, M., Crook, N. and Malaker, C.R. (1989). The Impact of Environment and Economic Class on Health in Urban India (case-studies of Pune and Durgapur). Project Report, School of Oriental and African Studies, University of London.

Bawa, V.K.(1984). Industrialization, Urbanization and Planning in the Pune Metropolitan Region—a study of Pimpri-Chinchwad, in Ballhatchet, K. and Taylor, D. (eds.) (1984).

Beaver, M.W. (1973). Population, Infant Mortality and Milk, *Population Studies*, 27:2.

Blomkvist, H. (1988). The Soft State: housing reform and state capacity in urban India. Ph.D. thesis, Uppsala University.

Bose, A. (1978). *India's Urbanization, 1901-2001.* New Delhi: Tata McGraw-Hill.

Boserup, E. (1981). *Population and Technology.* Oxford: Blackwell.

British Government, House of Commons (1924-25). *Reports (East India), Statement Exhibiting the Moral and Material Progress and Condition of India in 1923-24.* London: HMSO.

Byres, T.J. (1981). The New Technology, Class Formation and Class Action in the Indian Countryside, *Journal of Peasant Studies,* 8:4.

Cadman D., and Payne G. (eds.) (1990). *The Living City: towards a sustainable future*. London: Routledge.

Carr, J.C. and Taplin, W. (1962). *History of the British Steel Industry*. Oxford University Press: Oxford.

Cassen, R.H. (1978). *India. Population, Economy, Society*. London: Macmillan.

Char, S.V. (1979). Viability of Vizag Steel Plant, *Economic and Political Weekly*, 14:45.

Coale, A.J. (1972). *Growth and Structure of the Human Population*. Princeton: Princeton University Press.

Coale, A.J. and Hoover, E.M. (1958). *Population Growth and Economic Development in Low Income Countries*. Princeton: Princeton University Press.

Connell, J., Dasgupta, B., Laishley, R., and Lipton, M. (1976). *Migration from Rural Areas: the evidence from village studies*. Oxford: Oxford University Press.

Crook, N. and Dyson, T., (1982). Urbanization in India: results of the 1981 Census, *Population and Development Review*, 8:1.

Crook, N. and Malaker, C.R. (1988). Child Mortality and Environmental Differentials in a Newly Industrialised Indian City: Durgapur in West Bengal. Project Report, School of Oriental and African Studies, University of London.

Crook, N. and Malaker, C.R. (1989). Estimating Adult Mortality from Census Data on Widowhood: the case of West Bengal (manuscript).

Crook, N., Ramasubban, R. and Singh, B. (1991). A Multi-dimensional Approach to the Social Analysis of the Health Transition: case-study of Bombay, in Cleland, J. and Hill, A. (eds.) (1991). *The Health Transition: methods and measures*. Canberra: The Australian National University.

Dandekar, H.C. (1986). *Men to Bombay, Women at Home*. Ann Arbor: University of Michigan.

Datta, S.B. (1986). *Capital Accumulation and Workers' Struggle in Indian Industrialisation: the case of the Tata Iron and Steel Company 1910-1970*. Stockholm: Almqvist and Wiksell International.

Datta, A. (1968). Financing Municipal Services, *The Indian Journal of Public Adminstration*, 14:3.

Davis, K., and Golden, H.H. (1954). Urbanization and the Development of Pre-industrial Areas, *Economic Development and Cultural Change*, 3.

Deshpande, L.K. (1983). *Segmentation of Labour Market: a case study of Bombay*. Bombay: Orient Longman.

D'Souza, V.S. (1975). Scheduled Castes and Urbanization in Punjab: an explanation, *Sociological Bulletin*, 24:1.

Doyal, L. (1976). *Political Economy of Health*. London: Pluto.

Dyson, T., and Crook, N. (1984). *India's Demography: essays on the contemporary population.* New Delhi: South Asian Publishers.

Dyson, T. and Moore, M. (1983). On Kinship Structure, Female Autonomy, and Demographic Behaviour in India, *Population and Development Review,* 9:1.

Dyson, T. (ed.) (1989). *India's Historical Demography: studies in famine, disease, and society.* London: Curzon.

Glass, R. (1948). *The Social Background of a Plan: a study of Middlesbrough.* London: Routledge and Kegan Paul.

Gore, M.S. (1970). *Immigrants and Neighbourhoods: two aspects of life in a metropolitan city.* Bombay: Tata Institute of Social Sciences.

Gotpagar, K.B. (1990). Fertility Characteristics of Migrants and Non-migrants in Greater Bombay, in K. Srinivasan and K.B.Pathak (eds.), *Dynamics of Population and Family Welfare, 1989.* Bombay: Himalaya.

Government of India, Department of Health (1865). *First Annual Report of the Sanitary Commissioner for Bengal, 1864-65.* Calcutta: Military Orphan Press.

Government of India, Planning Commission (1956). *Second Five Year Plan.* Delhi: Manager of Publications.

Government of India, Ministry of Labour (1968). *Family Living Surveys among Industrial Workers, 1958-9.* Delhi: Manager of Publications.

Government of India, Ministry of Finance, Department of Economic Affairs (1972). *Tables with Notes on Internal Migration* (National Sample Survey No 182 (1964)). Delhi: Controller of Publications.

Government of India, Planning Commission (1972). *Report of the Working Group on Slums.* Delhi: Controller of Publications.

Government of India, Ministry of Work and Housing (1975). *Report of the Committee set up to review the Slum Clearance/Improvement Schemes for the Environmental Improvement in Slum Areas and other Matters.* Delhi: Controller of Publications.

Government of India, Registrar General (1976). *Fertility Differentials in India, 1972.* Delhi: Controller of Publications.

Government of India, Registrar General (1979). *Survey on Infant and Child Mortality, 1979.* Delhi: Controller of Publications.

Government of India, Registrar General (1981a). State of Madhya Pradesh, *Town Directory.* Delhi: Controller of Publications.

Government of India, Registrar General (1988). *Census of India, 1981, Occasional Papers No.5, Child Mortality Estimates of India.* Delhi: Controller of Publications.

Government of India, Ministry of Labour (various years). Labour Bureau, *Statistics of Factories.* Delhi: Controller of Publications.

Government of India, Registrar General (various years). *Census of India.* Delhi: Controller of Publications.

Government of India, Registrar General (various years). Sample Registration System *Bulletin*. Delhi: Controller of Publications.

Government of India, Sanitary Commissioner (various years). *Annual Report of the Sanitary Commissioner with the Government of India*. Calcutta: Superintendent of Printing.

Gugler, J. (ed.) (1988). *The Urbanization of the Third World*. Oxford: Oxford University Press.

Hardoy, J.E. and Satterthwaite, D. (1990). Urban Change in the Third World: are recent trends a useful pointer to the urban future?, in Cadman, D. and Payne, G. (eds.) (1990).

Harriss, J. (1984). Two Theses on Small Industry: notes from a study in Coimbatore, South India, in Ballhatchet, K. and Taylor, D. (eds.) (1984).

Harriss, J. (1989). Vulnerable Workers in the Indian Urban Labour Market, in Rodgers, G. (ed.) (1989). *Urban Poverty and the Labour Market: access to jobs and incomes in Asian and Latin American Cities*. Geneva: International Labour Organisation.

Holmstrom, M. (1976). *South Indian Factory Workers: their life and their world*. Cambridge: Cambridge University Press.

ILO-ARTEP (1989). *Employment and Structural Change in Indian Industries: a trade union view-point*. Geneva: ILO.

Jain, A.K. and Adlakha, A.L. (1984). Estimates of Fertility Decline in India during the 1970s, in Dyson, T. and Crook, N. (eds.) (1984).

Jaya Rao, K.S. (1985). Urban Nutrition in India, *Bulletin of the Nutrition Foundation of India*, 6:4.

Karve, I (1965). *Kinship Organization in India*. Bombay: Asia Publishing House.

Keenan, J.L. (1943). *A Steelman in India*. New York: Duell, Sloan and Pearce.

Kerkar, A. et. al. (1981). *Report of the High Power Steering Group for Slums and Dilapidated Houses*. Bombay: Government of Maharashtra.

Keynes, J.M. (1937). Some Economic Consequences of a declining Population, *Eugenics Review*, 29:1.

Klein, I. (1986). Urban Development and Death in Bombay City, 1870-1914, *Modern Asian Studies*, 20:4.

Lipton, M. (1977). *Why Poor People Stay Poor: urban bias in world development*. London: Maurice Temple Smith.

Mackie, J.W. (1982). Industrial Location and Regional Policy in South India. Ph.D. thesis, University of London.

Marx, K. (1967). *Capital*. Moscow: Progress Publishers.

Marx, K. (1983). The German Ideology (1844), in Marx, K and Engels, E. (1983), *Selected Works*. Moscow: Progress Publishers.

McNicoll, G. (1984). Consequences of Rapid Population Growth: an overview and assessment, *Population and Development Review*,

10:2.

Mera, K., (1973). On the Economic Agglomeration and Economic Efficiency, *Economic Development and Cultural Change*, 21:2.

Mills, E.S. and Becker, C.M. (1986). *Studies in Indian Urban Development* (World Bank Research Publication). Oxford: Oxford University Press.

Mills, E.S. (1972). *Urban Economics*. Glenview Illinois: Scott, Foresman and Company.

Mohan, R. and Pant, C. (1982). The Morphology of Urbanization in India: some results from the 1981 Census, *Economic and Political Weekly*, 17:39.

Mohsin, M. (1964). *Chittaranjan: a study in urban sociology*. Bombay: Popular Prakashan.

Moomaw, R.L. (1981). Productivity and City Size: a critique of the evidence, *Quarterly Journal of Economics*, 96.

Morris, M.D. (1960). The Recruitment of an Industrial Labour Force in India, with British and American Comparisons, *Comparative Studies in Society and History*, 2.

Morris, M.D. (1965). *The Emergence of an Industrial Labour Force in India: a study of the Bombay Cotton Mills 1854-1947*. Berkeley: University of California Press.

Municipal Commissioner of Bombay (various, dates). *Annual Report*. Bombay: Municipal Corporation of Greater Bombay.

Oberoi, A.S. and Singh, A.K.M. (1983). *Causes and Consequences of Internal Migration: a study in the Indian Punjab*. Delhi: Oxford University Press.

Prais, S.J. (1983). *Productivity and Industrial Structure*, (NIESR(UK) Economic and Social Studies xxxiii). Cambridge: Cambridge University Press.

Prakash, V. (1969). *New Towns in India*. Durham: Duke University Press.

Premi, M.K. (1980). Aspects of Female Migration in India, *Economic and Political Weekly*, 15:15.

Preston, S.H. (1979). Urban Growth in Developing Countries: a demographic re-appraisal, *Population and Development Review*, 5:2.

Psacharopoulos, G., and Woodhall, M. (1985). *Education for Development: an analysis of investment choices*. New York: Oxford University Press.

Pugh, C. (1989). *The World Bank and Housing Policy: the next steps*. University of Sheffield: Centre for Development Planning Studies.

Ramachandran, R. (1979). *Urbanization and Urban Systems in India*. Delhi: Oxford University Press.

Ray-Chaudhuri, J. (1990). Inter-urban and Rural-urban Linkages in Terms of Migration and Remittances: case-study in Durgapur, West Bengal. Ph.D. thesis, University of Cambridge.

Richardson, H.W., and Bass; J.M. (1965). The Profitability of Consett Iron Company before 1914, *Business History, 7:2*

Rothermund, D., et al. (eds.) (1980). *Urban Growth and Rural Stagnation: studies in the economy of an Indian coalfield and its rural hinterland.* New Delhi: Manohar.

Scherer, F.M. et al. (1975). *The Economics of Multi-Plant Operation.* Cambridge (Mass): Harvard University Press.

Sen, S.K. (1975). *The House of Tata (1839-1939).* Calcutta: Progressive Publishers.

Sengupta, N. (1985). Contract Labour in Rourkela Steel Plant, Madras Institute of Developing Studies Working Paper, No 57.

Shukla, V. (1988). *Urban Development and Regional Policy in India: an econometric analysis.* Bombay: Himalaya.

Sinha, A.P. (1989). Slum Development: cafeteria approach, *Urban India,* 9:2.

Sivaramakrishnan, K.C. (1976). *New Towns in India: a report on a study of selected new towns in the Eastern Region.* Calcutta: Indian Institute of Management.

Sivaramakrishnan, K.C. (1985). Case Study of an Indian Steel Town, in J.L.Taylor and D.G.Williams eds., *Urban Planning Practice in Developing Countries.* Oxford: Pergamon Press.

Skeldon, R. (1986). On Migration Patterns in India during the 1970s, *Population and Development Review,* 12:4.

Soni, V. (1984). The Development and Current Organisation of the Family Planning Programme, in T. Dyson, and N.Crook (eds.) (1984).

Sovani, N.V. (1964). The Analysis of 'Over-urbanisation', *Economic Development and Cultural Change,* 12.

Steel Authority of India Limited (1984). *History of Bhilai.* Bhilai: Bhilai Steel Plant.

Steel Authority of India Limited (formerly. Hindustan Steel Limited) (various years). *Iron and Steel: statistics for iron and steel industry in India.* Ranchi: Steel Authority of India Limited.

Times of India (various years). *Directory and Yearbook.* Bombay: Times of India Press.

Todaro, M.P. (1969). A Model of Labour Migration and Urban Unemployment in Less Developed Countries, *American Economic Review,* 59.

Turner, J.F.C. (1968). Housing Priorities, Settlement Patterns, and Urban Development in Modernising Countries, *Journal of American Institute of Planners,* 34:6.

United Nations (1989). *Housing and Economic Adjustment.* New York: Taylor and Francis.

Upadhyay, S.B. (1990). Cotton Mill Workers in Bombay, 1875 to 1918: conditions of work and life, *Economic and Political*

Weekly, 25:30.

Wadhva, K., (1983). Land Use Patterns in Urban Fringe Areas, *Economic and Political Weekly*, 18:14.

Weber, A.F. (1963). *The Growth of Cities in the Nineteenth Century: a study of statistics* (1899). Ithaca: Cornell University Press.

Weiner, M. (1978). *Sons of the Soil: migration and ethnic conflict in India*. Princeton: Princeton University Press.

Wildasin, D.E. (1986). *Urban Public Finance*. Chur: Harwood Academic Press.

Wilson, A.S. (1977). The Origins of the Consett Iron Company 1841-1864, *Durham University Journal*.

Woods, R.I., Waterson, P.A., and Woodward, J.H. (1989). The Causes of Rapid Infant Mortality Decline in England and Wales, 1861-1921, *Population Studies*, 42:3 and 43:1.

Wrigley, E.A. and Schofield, R.S. (1981). *The Population History of England, 1541-1871: a reconstruction*. London: Arnold.

NAME INDEX

SUBJECT INDEX